Nanofinishing of Materials for Advanced Industrial Applications

This book details nanofinishing techniques employed for finishing of industrial applications in a comprehensive manner. It provides an introduction into nanofinishing requirements and basic nanofinishing techniques and their variants and details the latest developments, research and innovation, and advancements. It includes topics ranging from working principles, material removal mechanisms, force analysis, and mathematics involved in industry-specific applications, salient features, comparative study, and automation and optimization in nanofinishing technologies.

This book:

- Explains cutting-edge nanofinishing technologies comprehensively,
- Discusses numerical explanations for better selection of process parameters and their modelling and optimization,
- Provides application-oriented aspects of nanofinishing processes,
- Explains processes considering Industry 4.0 capabilities and green manufacturing,
- Focuses on applications with good relevance to industry and coverage of case studies.

This book is aimed at graduate students and researchers in manufacturing and process engineering, and the die and mould industries.

Advanced Materials Processing and Manufacturing

Series Editor: Kapil Gupta

The CRC Press Series in Advanced Materials Processing and Manufacturing covers the complete spectrum of materials and manufacturing technology, including fundamental principles, theoretical background, and advancements. Considering the accelerated importance of advances for producing quality products for a wide range of applications, the titles in this series reflect the state-of-the-art in understanding and engineering the materials processing and manufacturing operations. Technological advancements for enhancement of product quality, process productivity, and sustainability, are on special focus including processing for all materials and novel processes. This series aims to foster knowledge enrichment on conventional and modern machining processes. Micro-manufacturing technologies such as micro-machining, micro-forming, and micro-joining, and Hybrid manufacturing, additive manufacturing, near net shape manufacturing, and ultra-precision finishing techniques are also covered.

Advanced Materials Characterization
Basic Principles, Novel Applications, and Future Directions
Ch Sateesh Kumar, M. Muralidhar Singh and Ram Krishna

Thin-Films for Machining Difficult-to-Cut Materials
Challenges, Applications, and Future Prospects
Ch Sateesh Kumar and Filipe Daniel Fernandes

Advanced Materials Processing and Manufacturing
Research, Technology, and Applications
Amogelang Sylvester Bolokang and Maria Ntsoaki Mathabathe

Advanced Joining Technologies
Edited by Manjaiah M, Shivraman Thapliyal and Adepu Kumar

Nanofinishing of Materials for Advanced Industrial Applications
Edited by Faiz Iqbal, Dilshad Ahmad Khan, and Zafar Alam

For more information about this series, please visit: www.routledge.com/Advanced-Materials-Processing-and-Manufacturing/book-series/CRCAMPM

Nanofinishing of Materials for Advanced Industrial Applications

Edited by
Faiz Iqbal, Dilshad Ahmad Khan,
and Zafar Alam

CRC Press
Taylor & Francis Group
Boca Raton London New York

CRC Press is an imprint of the
Taylor & Francis Group, an **informa** business

Designed cover image: Shutterstock

First edition published 2025
by CRC Press
2385 NW Executive Center Drive, Suite 320, Boca Raton FL 33431

and by CRC Press
4 Park Square, Milton Park, Abingdon, Oxon, OX14 4RN

CRC Press is an imprint of Taylor & Francis Group, LLC

ISBN: 9781032527963 (hbk)
ISBN: 9781032802992 (pbk)
ISBN: 9781003496298 (ebk)

DOI: 10.1201/9781003496298

Typeset in Times
by codeMantra

Contents

Preface

In a world where precision and sustainability are paramount in the industrial landscape, the art and science of nanofinishing have emerged as a critical frontier. The quest for achieving optimal surface quality, improved functionality, and eco-friendly practices has fuelled research and innovation in the field of nanofinishing. As materials scientists and engineers, we stand at the crossroads of opportunity, where the meticulous refinement of surfaces offers the potential to transform industries and impact our everyday lives.

This edited book ***Nanofinishing of Materials for Advanced Industrial Applications*** is the culmination of the collective efforts of researchers and experts who have dedicated themselves to unravelling the intricacies of nanofinishing. Through their contributions, this volume explores the multifaceted world of nanofinishing, encompassing the realms of optimization and sustainability, the intricacies of rotational magnetorheological abrasive flow finishing, the art of nanofinishing metallic mirrors, and the extraordinary advances in biomedical implant applications in healthcare.

The journey begins with an in-depth exploration of sustainability, emphasizing the optimization of resources and nanofinishing parameters, setting the stage for a sustainable industrial future. The chapters delve into the world of multi-objective optimization, revealing cutting-edge developments in rotational magnetorheological abrasive flow finishing processes, demonstrating how the synergy of science and engineering can produce superior outcomes.

The significance of nanofinishing becomes vividly apparent as we venture into the realms of metallic mirrors, biomedical implants, and the role of surface topography in promoting osseointegration and antibacterial properties and discover how nanofinishing techniques contribute to the art and science of materials science and engineering, as well as the biomedical field.

With a focus on innovation and precision, this volume offers insights into material removal mechanisms during magnetorheological finishing and reviews advanced finishing processes for internal cylindrical surfaces. It also delves into the crucial post-processing methods for enhancing the surfaces of additively manufactured parts, paving the way for the industries of tomorrow.

Automation, too, takes centre stage as we explore the automation of abrasive-based precision-finishing techniques. We uncover the advantages and limitations of automation and how it interfaces with nanofinishing, promising a future marked by efficiency and precision.

In this book, we aim to provide a comprehensive and interdisciplinary view of nanofinishing techniques, their applications, and their potential to redefine the way we approach industrial processes. We invite you to embark on a journey through the intricacies of nanofinishing and join us in exploring the boundless opportunities that this field offers. This book serves as a resource for researchers, engineers, and students eager to contribute to the advancement of industrial applications through nanofinishing techniques.

We extend our gratitude to the authors, whose expertise and dedication have brought this book to fruition. Their collective knowledge and insights are invaluable, and their commitment to pushing the boundaries of nanofinishing is evident in the depth of content presented here.

May this book inspire, educate, and guide you on your own journey into the captivating world of nanofinishing, where precision, sustainability, and innovation converge to shape the industries of tomorrow.

Editors: Faiz Iqbal, Dilshad Ahmad Khan, Zafar Alam

Editors' biographical notes

Dr Faiz Iqbal is a Mechanical Engineering Lecturer at the University of Lincoln. Previously, he served as an Assistant Professor at Mahindra Ecole Central School of Engineering in India, focusing on Mechatronics, Industry 4.0, and Mechanical Engineering. He gained experience as a Post-Doctoral Research Associate at the University of Edinburgh, specializing in robotics for extreme environments where he co-founded Fluidic Logic Limited, a spin-off company from his entrepreneurial endeavours. Additionally, he secured EPSRC funding for COVID-19 research, aiding a Scottish micro-SME in scaling up disinfectant production using Industry 4.0 technologies.

He received his bachelor's degree in Mechanical and Automation engineering from Maharshi Dayanand University, India, in 2011, his master's degree in Mechatronics Engineering from Amity University, India, in 2013, and a PhD in Manufacturing Automation from the Indian Institute of Technology Delhi, India, in 2019. He is currently working as a Lecturer in the School of Engineering, University of Lincoln, Lincoln, United Kingdom. As an academician and researcher, he has to his credit several research papers in peer-reviewed international journals and conferences. His research interests include but are not limited to Mechatronics; Industrial Automation; Smart Manufacturing; Fluidic Logic/Soft Robotics; Advanced Machining Processes; Non-Conventional Machining; Industry 4.0—Cyber Physical Systems; Advanced Metrology.

Dr Dilshad Ahmad Khan is an Assistant Professor at the Department of Mechanical Engineering, NIT, Hamirpur, India. He is a prominent academician and researcher. He is a Chartered Engineer (CEng) from the Institution of Mechanical Engineers, London. He has received his PhD in Manufacturing from the Indian Institute of Technology (IIT) Delhi, India, in 2018 and Master of Technology from Aligarh Muslim University, India, in 2010. He has received his Bachelor of Engineering in Mechanical Engineering from Dr B.R. Ambedkar University (formerly Agra University), India, in 2006. He has published a good number of research articles in international journals and conferences. He has contributed several book chapters on various topics and has filed seven Indian patents. This book is the author's third published book and there are additional books in the process of being published. He has received various national and international awards for his research and innovations. His research interests include advanced finishing/polishing processes, non-conventional machining, additive manufacturing, mechatronic systems, and industrial automation.

Dr Zafar Alam serves as an Assistant Professor in the Department of Mechanical Engineering at the Indian Institute of Technology (IIT; Indian School of Mines (ISM)) in Dhanbad, India. Prior to his tenure at IIT (ISM) Dhanbad, he worked as an Assistant Professor of Mechanical Engineering at Zakir Husain College of Engineering and Technology, Aligarh Muslim University, India, for a duration of two years. His educational background includes a B. Tech in Mechanical Engineering obtained from Jamia Millia Islamia, India, in 2012, as well as an M.Tech and a PhD in Production Engineering, both acquired from IIT, Delhi, India, in 2014 and 2019, respectively.

Dr Zafar Alam is a prolific academic and researcher with a substantial body of work to his name, including an authored book, two edited books, numerous book chapters, and over two dozen research papers published in peer-reviewed international journals and conferences. His achievements also encompass six Indian patents and the receipt of two international and three national awards, notably the prestigious GYTI (Gandhian Young Technological Innovation) award, in recognition of his outstanding contributions to the realm of research and innovation.

His research interests encompass various areas, including advanced finishing and polishing processes, non-conventional machining, additive manufacturing, industrial automation, and motion control, among others.

Contributors

Shadab Ahmad
School of Mechanical Engineering, Shandong University of Technology, Zibo, Shandong, China

Zafar Alam
Department of Mechanical Engineering, Indian Institute of Technology (Indian School of Mines), Dhanbad, Jharkhand, India

Parvesh Ali
Gautam Buddha University, Greater Noida, Uttar Pradesh, India

Navneet Arora
Department of Mechanical and Industrial Engineering, Indian Institute of Technology, Roorkee, India

Anant Bhardwaj
Department of Mechanical Engineering, Poornima College of Engineering, Jaipur, India

Abhishek Bhatnagar
Department of Mechanical Engineering, Indian Institute of Technology Delhi, New Delhi, India

Anirudh Chana
Department of Mechanical Engineering, Thapar Institute of Engineering and Technology, Patiala, India

Ajay Choudhary
Department of Mechanical Engineering, National Institute of Technology Hamirpur, Himachal Pradesh, India

Manas Das
Indian Institute of Technology Guwahati, Guwahati, India

Kalyanmoy Deb
Electrical and Computer Engineering, Michigan State University, East Lansing, MI, USA

Manojit Ghosh
Department of Metallurgy and Materials Engineering, Indian Institute of Engineering Science and Technology, Shibpur, West Bengal, India

Pushpendra Gupta
Department of Mechanical Engineering, Indian Institute of Technology, Kharagpur, West Bengal, India

Abdul Wahab Hashmi
School of Engineering, University of Lincoln, Lincoln, LN6 7TS, United Kingdom

Faiz Iqbal
School of Engineering, University of Lincoln, Lincoln, LN6 7TS, United Kingdom

Varnit Jain
Department of Metallurgy and Materials Engineering, Indian Institute of Engineering Science and Technology, Shibpur, West Bengal, India

Sajan Kapil
Indian Institute of Technology Guwahati, Guwahati, India

Dilshad Ahmad Khan
Department of Mechanical Engineering, National Institute of Technology, Hamirpur, Himachal Pradesh, India

Vidyapati Kumar
Department of Mechanical Engineering, Indian Institute of Technology, Kharagpur, West Bengal, India

Munish Kumar
Department of Mechanical Engineering, National Institute of Technology, Hamirpur, Himachal Pradesh, India

Manish Kumar
Department of Mechanical Engineering, Indian Institute of Technology (Indian School of Mines), Dhanbad, Jharkhand, India

Shrushti Maheshwari
Department of Mechanical Engineering, Indian Institute of Technology (Indian School of Mines), Dhanbad, Jharkhand, India

Ankita Mistri
Department of Mechanical Engineering, Indian Institute of Technology, Dhanbad, Jharkhand, India

Mohammed Omar Abdullah Ba Matraf
School of Mechanical Engineering, Shandong University of Technology, Zibo, Shandong, China

Sunil Kumar Paswan
School of Engineering, Design and Automation, GNA University, Sri Hargobindgarh, Phagwara, Punjab, India

Deepak Patil
Department of Production Engineering, National Institute of Technology, Tiruchirappalli, Tamil Nadu, India

Dilip Kumar Pratihar
Department of Mechanical Engineering, Indian Institute of Technology, Kharagpur, West Bengal, India

Atul Singh Rajput
Indian Institute of Technology Guwahati, Guwahati, India

Saurabh Singh Rathore
Department of Mechanical and Industrial Engineering, Indian Institute of Technology, Roorkee, India

Varun Sharma
Department of Mechanical and Industrial Engineering, Indian Institute of Technology, Roorkee, India

Anant Kumar Singh
Department of Mechanical Engineering, Thapar Institute of Engineering and Technology, Patiala, India

Yebing Tian
School of Mechanical Engineering, Shandong University of Technology, Zibo, Shandong, China

Rajan Kumar Verma
Department of Metallurgy and Materials Engineering, Indian Institute of Engineering Science and Technology, Shibpur, West Bengal, India

1 Nanofinishing in light of optimization and sustainability

Abhishek Bhatnagar, Zafar Alam, Dilshad Ahmad Khan, and Faiz Iqbal

1.1 INTRODUCTION

The aim of this chapter is to understand the concept of sustainability and sustainable development that helps growth and development while not compromising on the future. This requires judicious use of natural resources, which implies optimizing a certain objective while keeping the available resources for the future as well. This chapter presents several definitions of sustainability and sustainable development, and closely links them with optimization. Further, both these concepts are applied to manufacturing processes (nanofinishing in particular) to understand the optimization of machining parameters. The optimal values of these parameters help in the designing of sustainable manufacturing systems that can have throughputs to cater to the needs of people while not compromising on the future.

The chapter is organized into three broad sections and the conclusion. The first section (Section 1.2) details out the concepts of sustainability, sustainable development, and enablers and barriers to sustainability. The second section (Section 1.3) provides on overview of the broad field of optimization with a few key methods and examples for optimization of resources. The third section (Section 1.4) provides the optimization of parameters for a nanofinishing process. Section 1.5 concludes the chapter.

1.2 SUSTAINABLE DEVELOPMENT AND SUSTAINABILITY

Sustainable development goals (SDGs) are adopted by the United Nations for a better and sustainable future for all [1, 2]. These refer to a set of 17 inter-related goals whose broad themes under SDGs include dignity (SDGs 1, 5), people (SDGs 2, 3, 4), planet (SDGs 6, 12, 13, 14, 15), partnership (SDG 17), justice (SDG 16), and prosperity (SDGs 7, 8, 9, 10, 11) [3, 4]. These goals aim to improve the standard of living broadly through better education, environmental consciousness, poverty reduction, equality, good health, sustainable cities, and communities. However, improved standard of living requires development in all sectors and this development aspect has its own challenges. Development needs to be carried out in a way that humans co-exist on

DOI: 10.1201/9781003496298-1

the planet over a long duration, thus keeping in view the environment while carrying out development-related activities. This leads to an important concept of sustainable development, which is main idea in this discourse. This has various associated interpretations based on the context of its usage. A few definitions of sustainable development are given next:

1. Schaefer and Crane [5] define sustainable development as "the type of development process that meets the needs of the current generation without compromising the ability of future generations to meet their own needs." This definition implies judicious use of non-renewable resources by the current generation to make it available for the future as well. Development should continue keeping in view sustainability for all.
2. Mensah [6] defines sustainable development as "a mechanism through which society can interact with the environment while not risking damaging the resource for the future." This essentially implies improving the quality of life without endangering the ecosystems and the environment in general that may cause climate change, pollution of any form (air, noise, and water), deforestation, and extinction of wildlife and birds [7].
3. Zhai and Chang [8] define sustainable development as a development paradigm that "aims at achieving social progress, environmental equilibrium and economic growth." This implies meeting the developmental goals and simultaneously sustaining the ability of systems that provide the non-renewable resources and do not hamper the social and economic ecosystem.

The common ideas that come through these definitions lead to incorporation of sustainability concepts in different spheres of life while developing various systems for the future. A key part of development lies in manufacturing of goods using various tools and techniques. These goods are used by people to meet their daily requirements; hence, they require large-scale production units with high throughput, enhanced capacities, and precision for customer satisfaction over the long run. The large scale of manufacturing requires these processes to be environment friendly and economical, as well as meet the requirements of the people. In short, the manufacturing processes should be sustainable in the long run and the concept of sustainability needs to be included in these processes.

Sustainability has several definitions in the literature; however, the keywords used in all definition are people, costs, environment, holistic, future, and long run. A few definitions are as follows:

1. Sustainability can be defined as "the capacity to maintain or improve the state and availability of desirable materials or conditions over the long term" [9]. This definition emphasizes on the long-term impact that requires either to maintain the system at a state where it can sustain easily or to improve it incrementally (since, there is no endpoint) to adopt new technologies that can be helpful to cater to the needs of the future.
2. Sustainability is "a dynamic equilibrium in the process of interaction between the population and the carrying capacity of its environment such

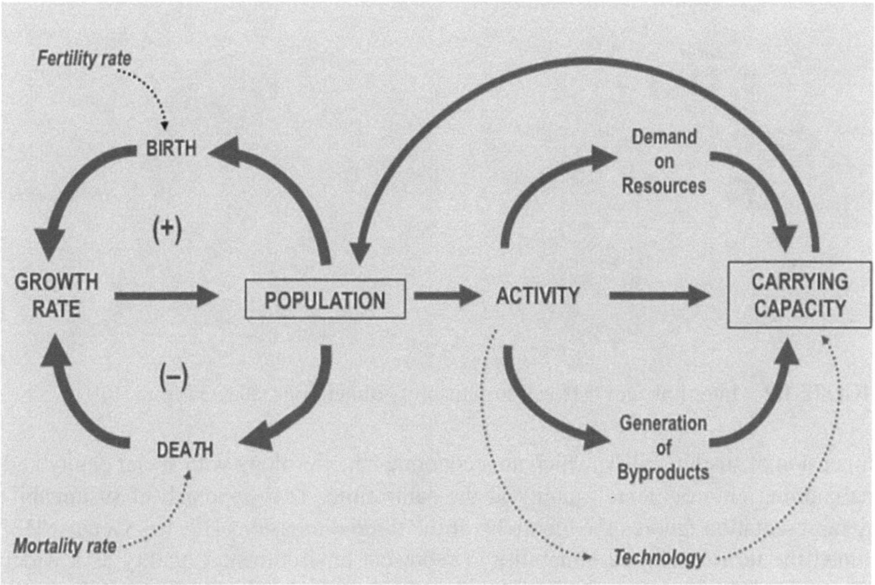

FIGURE 1.1 Population and its activity; adapted from [10].

that the population develops to express its full potential without producing irreversible adverse effects on the carrying capacity of the environment upon which it depends" [10]. This definition provides a holistic view of the planet, its species, its future, and the interaction between the population and its carrying capacity. Both these factors affect each other as shown in Figure 1.1.

3. Sustainability is "our ability to maintain a coupled human–nature system at a desirable state for multiple generations in the face of anthropogenic and environmental perturbations and uncertainties" [11]. This definition stresses upon the continuous interaction between humans and nature and the effect that one has on the other.

   As evident from the definitions stated above, sustainability essentially means to sustain for the future, which gives rise to three pillars of sustainability. These pillars are economy, society, and the environment. Subsequently, these give rise to three dimensions of sustainability, namely, economic sustainability, social sustainability, and environmental sustainability. These three dimensions are interlinked using different concepts as given in Figure 1.2, as adapted from [16] and elaborated later.

Figure 1.2 demonstrates sustainability from different perspectives, namely, triple bottom line, weak sustainability, and strong sustainability. The triple bottom line concept of sustainability refers to the working of an organization towards achieving their goals while keeping the views of people, the planet, and profits into consideration. This is typically represented by a Venn diagram whose intersection represents sustainability. This aims to cover the long-term strategies of an organization towards achieving the

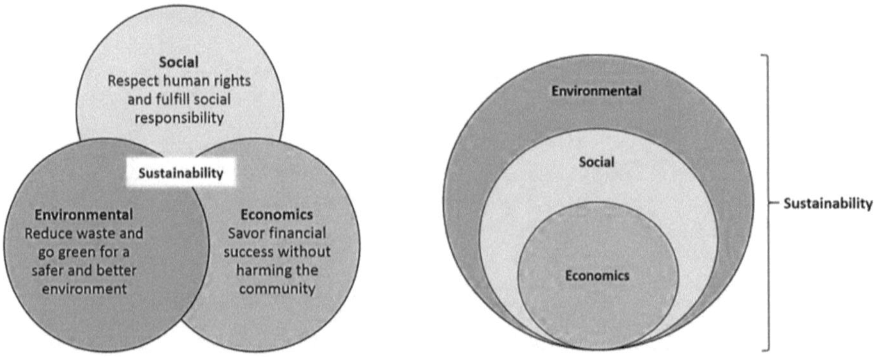

FIGURE 1.2 Inter-linkages between sustainability dimensions; adapted from [16].

dimension of sustainability, which are economic benefits along with social equity and maintaining environmental quality at the same time. This approach of sustainability representation ignores the hierarchy of the three dimensions [12, 13]. Getzner [13] coined the term strong sustainability to represent environmental quality as a wider perspective in the system with social aspects in the next level and economic factor as the last part of this hierarchy. This means that economic development requires both social and environmental capital; and the environment provides the required resources of social and economic development. It implies that the natural capital (environmental systems) and the man-made capital (society and economy) are not substitutable. The increase in population and its negative effects on the environmental systems due to scarcity of natural resources require an emphasis on the environmental dimension, putting it at the highest level in the context of strong sustainability. The weak sustainability aspect stresses the fact that overall sustainability remains constant while the individual dimensions can be substituted. The three dimensions of sustainability have sub-areas within which they operate as per Helleno et al. [14]. The economic dimension includes cost management, corporative management, operational efficiency, inclusion of new products, operating results, suppliers, customers, and infrastructure. The social dimension comprises employee salary and benefits, their satisfaction levels, quality and safety, human resources, and the community. The environmental dimension includes environmental aspects and their management, responsible consumption, and life cycle of the product and its recycling [15].

The manufacturing of goods comprises several processes throughout their life cycle while being manufactured and up to the time they are used by the customer after which they are recycled and remanufactured. This cycle is represented in Figure 1.3. Incorporation of sustainability concepts in this life cycle is important due to the large scale of production. This requires working as per the sustainable aggregate production plans as illustrated in Figure 1.4, which indicates maximization of total profits, focus on the use of renewable resources (reduction in usage of non-renewable resources), satisfaction of employees, and focus on their well-being.

A sustainable machining process has several components, as illustrated in Figure 1.5. It shows the intersection of the sustainability dimensions during the machining process. The economic dimensions focus on consumption of energy that

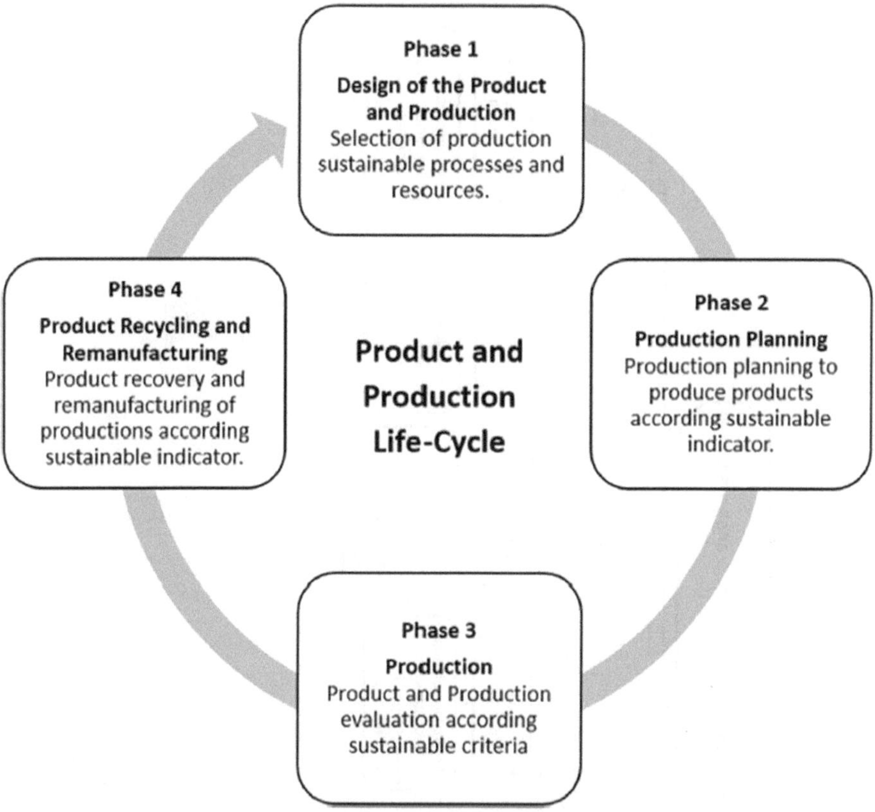

FIGURE 1.3 Product and production life cycle; adapted from [16].

FIGURE 1.4 Pillars of sustainable aggregate production plans; adapted from [17].

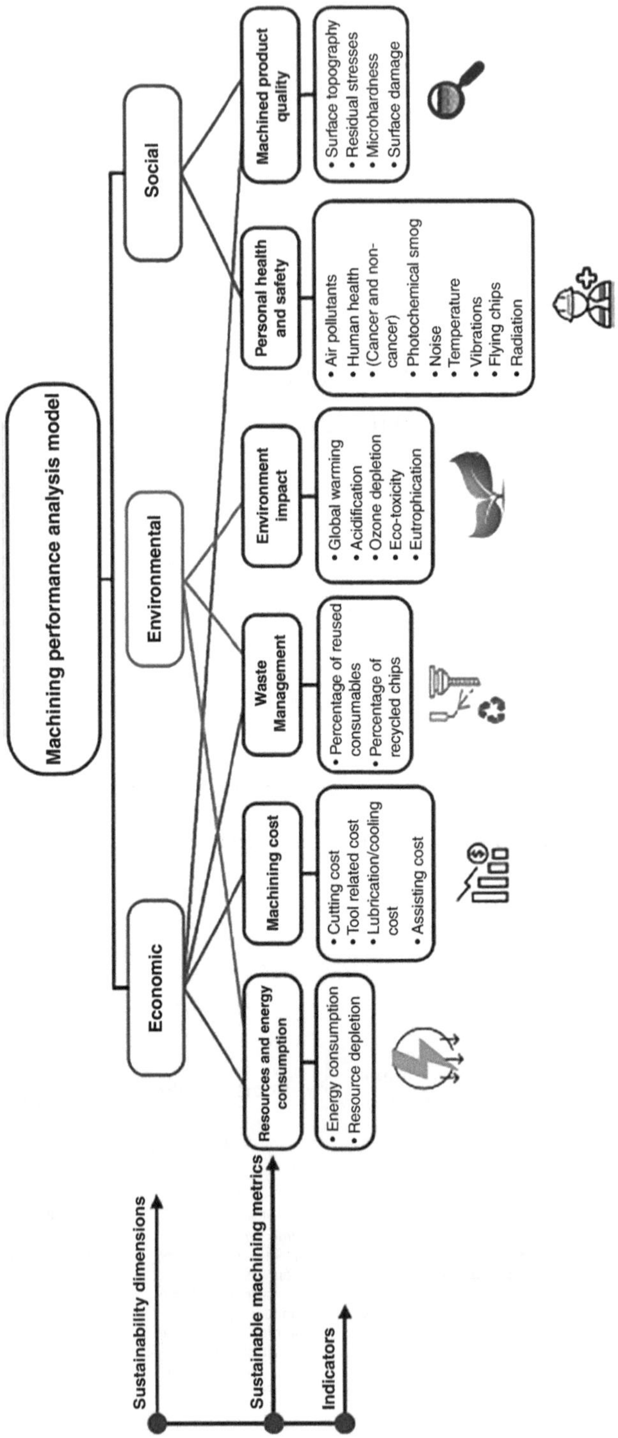

FIGURE 1.5 Sustainable machining process analysis; adapted from [19].

leads to resource depletion, fixed and variable costs involved in the machining process such as tool costs, lubrication costs, etc. Waste management becomes a critical part that involves cost that is a function of reusability and recycling. The machined product goes into the market which should be profitable for the organization and its profitability depends on its quality of finishing as well as the production process in general. The environmental aspects include the resource depletion aspect, management of waste, and the impact on the environment such as global warming and climate change. The social aspects include the health and safety of individuals involved in the production process and the product quality for its enhanced consumption in the market [18].

Given the impact of the large-scale production and machining processes, it is imperative to include optimization aspects in this process from the procurement of raw materials to the delivery to the final consumers, as given in the next section.

1.3 OPTIMIZATION OF RESOURCES

Section 1.2 presented the concepts of sustainable development and sustainability, and its needs in the large-scale production processes. In this and the next section, we will discuss how the optimization of resources (better usage of manpower, infrastructure, and transportation) and the production processes can help achieve SDGs. Optimization is an act of making decisions using the available resources which can be constraints given a limited budget. It is a broad field with wide applications, which is required for sustainable operations. An optimization problem is characterized by decision variables which are the output of the model. The values of these variables optimize a certain objective. The model consists of an objective function; constraints that refer to the considerations to be incorporated while optimizing the objective; and parameters whose values are varied in different experiments. An optimization problem is mathematically modelled and solved using different solution techniques that depend on its mathematical complexity and nature of the formulation [20, 21].

Optimization has various use cases in the manufacturing setup. We will discuss a few applications of optimization in the entire supply chain from the procurement of raw materials to the delivery till the final consumer. Large-scale production requires setting up the manufacturing unit at an appropriate location. This is known as a facility location problem [22], whose decision variables represent where to locate raw material outlets, the manufacturing location given that the customer locations are known. In the presence of many manufacturing units and raw material suppliers, it is required to map these suppliers to the manufacturing units optimally using the assignment models and vendor selection models [23]. Next, the routing of raw material vehicles from suppliers to the manufacturing plant and the manufacturing plant to distribution centres and then to the customers takes place. These are through vehicle routing models [24]. The layout of the manufacturing plant to optimize the throughput when an item undergoes the transformation from raw material to a finished product happens through facility layout models [25]. Further, the processes through which these operations take place can also be optimized, that is, different parameters of the machining process can be adjusted to use the resources judiciously given the large scale of production. These are discussed in detail specifically for the nanofinishing process in the next section.

The optimization problems are formulated as mathematical models that can be solved to get the values of the decision variables. The solution methods include exact and heuristic methods that can be applied on standard solvers to get the output. The exact method gives the optimal solution but takes a long time as compared to the heuristic methods which are sub-optimal due to lower computational time. The commonly used heuristic methods include genetic algorithm, ant colony optimization, artificial bee colony optimization, simulated annealing, and tabu search, etc. The next section discusses the optimization of machining parameters for the nanofinishing process.

1.4 OPTIMIZATION OF NANOFINISHING PARAMETERS

Optimization of nanofinishing parameters is a crucial aspect for improving surface quality and performance in various industries. It involves fine-tuning a range of factors to achieve the desired outcomes, such as surface roughness reduction, increased wear resistance, and improved functional properties. Here is a brief literature review highlighting key studies and approaches in the optimization of nanofinishing parameters:

Optimization Techniques: Researchers have employed various optimization techniques, including Taguchi methods, response surface methodology (RSM), and genetic algorithms, to determine the optimal combination of parameters such as feed rate, tool speed, pressure, and abrasive size. These methods help identify the settings that yield the best surface finish and minimize defects.

Material-Specific Optimization: Different materials require tailored approaches. Studies have focused on optimizing nanofinishing parameters for metals [26–28], ceramics [29], polymers [30], and composites [31]. For example, in the case of metals, optimizing parameters like electrolyte composition in electrochemical polishing can significantly enhance surface quality.

Multi-Objective Optimization: In many applications, achieving multiple objectives is essential. Researchers have employed multi-objective optimization techniques to balance conflicting goals, such as minimizing surface roughness while maximizing material removal rates [32].

Tooling and Abrasives Optimization: Selecting the right tools and abrasives is vital. Investigations into the selection of abrasive materials, abrasive size, and tool geometry have been conducted to optimize nanofinishing processes [33–35].

Environmental Considerations: Sustainability is becoming increasingly important in manufacturing. Researchers are optimizing parameters with a focus on reducing energy consumption and minimizing waste in nanofinishing processes [36].

In-Process Monitoring and Control: Real-time monitoring and control systems have been integrated into nanofinishing processes to adapt and optimize parameters as conditions change. This ensures consistent quality and efficiency [37–40].

Surface Characterization Techniques: Alongside parameter optimization, researchers have employed advanced surface characterization techniques like atomic force microscopy (AFM) and scanning electron microscopy (SEM) to assess the effects of different parameter settings on surface properties.

Applications: Optimization of nanofinishing parameters has been applied across various industries, including automotive, aerospace, electronics, and medical devices, with a focus on enhancing product performance and reducing manufacturing costs.

The optimization of nanofinishing parameters is a dynamic field that plays a crucial role in achieving superior surface quality and sustainability across a range of industries. Researchers continue to explore innovative approaches and technologies to enhance the efficiency and effectiveness of nanofinishing processes.

1.5 CONCLUSION

The introductory chapter to this book underscores the paramount significance of optimizing nanofinishing parameters while concurrently prioritizing sustainability. This convergence of precision and eco-consciousness holds immense promise for industries seeking to enhance product quality and reduce their environmental footprint. The chapter highlights the multifaceted nature of nanofinishing parameter optimization, encompassing techniques, materials, multi-objective goals, and tools. Moreover, the chapter underpins the increasing integration of real-time monitoring and control systems, coupled with advanced surface characterization techniques, to refine nanofinishing processes. Crucially, the pursuit of sustainability is intrinsic to contemporary manufacturing practices. Nanofinishing optimization offers a unique avenue to align production objectives with environmental responsibility. It can significantly reduce material waste, energy consumption, and adverse ecological impacts. The chapter underscores the synergy between green engineering principles, circular economy concepts, and nanofinishing, fostering a sustainable paradigm.

This book further explains the advancements in nanofinishing of materials for advanced industrial applications spread over ten chapters including this one. Following on from this chapter, Chapter 2 explores the NSGA-II optimization technique for nanofinishing of a β-phase Ti-Nb-Ta-Zr alloy, Chapter 3 explores nanofinishing of metallic mirrors, Chapter 4 analyses nanofinishing of biomedical implants, Chapter 5 looks at the role of nano-micro surface topography in biomedical applications, Chapter 6 investigates the material removal mechanism for industrial extrusion punch surface, Chapter 7 provides insights into nanofinishing of internal cylindrical surfaces, Chapter 8 covers nanofinishing of 3D printed parts, which is another rapidly emerging industrial application, Chapter 9 then provides an overview of automation in nanofinishing techniques, and lastly Chapter 10 discusses the latest innovations in precision finishing techniques.

REFERENCES

1. Mio, C., Panfilo, S., & Blundo, B. (2020). Sustainable development goals and the strategic role of business: A systematic literature review. *Business Strategy and the Environment*, 29(8), 3220–3245.
2. Orhan, C. C., & Guajardo, M. (2022). Analytics in developing countries: Methods, applications, and the impact on the UN Sustainable Development Goals. *International Transactions in Operational Research*, 29(4), 2041–2081.
3. Bhatnagar, A., & Bolia, N. B. (2023). A sustainable decision-making framework for school consolidation policy. *Regional Science Policy & Practice*, 15(5), 1037–1063.
4. Leal Filho, W., Azeiteiro, U., Alves, F., Pace, P., Mifsud, M., Brandli, L., ... & Disterheft, A. (2018). Reinvigorating the sustainable development research agenda: The role of the sustainable development goals (SDG). *International Journal of Sustainable Development & World Ecology*, 25(2), 131–142.
5. Schaefer, A., & Crane, A. (2005). Addressing sustainability and consumption. *Journal of Macromarketing*, 25(1), 76–92.
6. Mensah, J. (2019). Sustainable development: Meaning, history, principles, pillars, and implications for human action: Literature review. *Cogent Social Sciences*, 5(1), 1653531.
7. Browning, M. H., & Rigolon, A. (2019). School green space and its impact on academic performance: A systematic literature review. *International Journal of Environmental Research and Public Health*, 16(3), 429.
8. Zhai, T., & Chang, Y. C. (2018). Standing of environmental public-interest litigants in China: Evolution, obstacles, and solutions. *Journal of Environmental Law*, 30(3), 369–397.
9. Harrington, L. M. B. (2016). Sustainability theory and conceptual considerations: A review of key ideas for sustainability, and the rural context. *Papers in Applied Geography*, 2(4), 365–382.
10. Ben-Eli, M. U. (2018). Sustainability: Definition and five core principles, a systems perspective. *Sustainability Science*, 13(5), 1337–1343.
11. Wu, J., & Wu, T. (2012). Sustainability indicators and indices: An overview. *Handbook of Sustainability Management*, 65–86.
12. Correia, M. S. (2019). Sustainability: An overview of the triple bottom line and sustainability implementation. *International Journal of Strategic Engineering (IJoSE)*, 2(1), 29–38.
13. Getzner, M. (1999). Weak and strong sustainability indicators and regional environmental resources. *Environmental Management and Health*, 10(3), 170–177.
14. Helleno, A. L., de Moraes, A. J. I., & Simon, A. T. (2017). Integrating sustainability indicators and Lean Manufacturing to assess manufacturing processes: Application case studies in Brazilian industry. *Journal of Cleaner Production*, 153, 405–416.
15. Akbar, M., & Irohara, T. (2018). Scheduling for sustainable manufacturing: A review. *Journal of Cleaner Production*, 205, 866–883.
16. Zarte, M., Pechmann, A., & Nunes, I. L. (2019). Decision support systems for sustainable manufacturing surrounding the product and production life cycle–A literature review. *Journal of Cleaner Production*, 219, 336–349.
17. Rasmi, S. A. B., Kazan, C., & Türkay, M. (2019). A multi-criteria decision analysis to include environmental, social, and cultural issues in the sustainable aggregate production plans. *Computers & Industrial Engineering*, 132, 348–360.
18. Joung, C. B., Carrell, J., Sarkar, P., & Feng, S. C. (2013). Categorization of indicators for sustainable manufacturing. *Ecological Indicators*, 24, 148–157.
19. Salem, A., Hegab, H., & Kishawy, H. A. (2021). An integrated approach for sustainable machining processes: Assessment, performance analysis, and optimization. *Sustainable Production and Consumption*, 25, 450–470.

20. Bhatnagar, A., Gupta, A., Joshi, A., & Bolia, N. (2022). An integrated framework for the improvement of school bus services: Understanding commuters' perceptions for sustainable school bus transportation. *Habitat International*, 126, 102602.
21. Bhatnagar, A., & Bolia, N. B. (2019). Improved governance of Indian school system through school consolidation. *Journal of Policy Modeling*, 41(6), 1160–1178.
22. Sachdeva, A., Singh, B., Prasad, R., Goel, N., Mondal, R., Munjal, J., ... & Dahiya, M. (2022). Metaheuristic for hub-spoke facility location problem: Application to Indian e-commerce industry. arXiv preprint arXiv:2212.08299.
23. Deretarla, Ö., Erdebilli, B., & Gündoğan, M. (2023). An integrated Analytic Hierarchy Process and Complex Proportional Assessment for vendor selection in supply chain management. *Decision Analytics Journal*, 6, 100155.
24. Konstantakopoulos, G. D., Gayialis, S. P., & Kechagias, E. P. (2020). Vehicle routing problem and related algorithms for logistics distribution: A literature review and classification. *Operational Research*, 22(3), 2033–2062.
25. Subulan, K., Varol, B., & Baykasoğlu, A. (2023). Unequal-area capability-based facility layout design problem with a heuristic decomposition-based iterative mathematical programming approach. *Expert Systems with Applications*, 214, 119199.
26. Iqbal, F., & Jha, S. (2019). Experimental investigations into transient roughness reduction in ball-end magneto-rheological finishing process. *Materials and Manufacturing Processes*, 34(2), 224–231.
27. Alam, Z., Khan, D. A., & Jha, S. (2019). MR fluid-based novel finishing process for nonplanar copper mirrors. *The International Journal of Advanced Manufacturing Technology*, 101, 995–1006.
28. Khan, D. A., Alam, Z., Iqbal, F., & Jha, S. (2017, July). A study on the effect of polishing fluid composition in ball end magnetorheological finishing of aluminum. *In 39th International MATADOR Conference on Advanced Manufacturing*. UK: University of Manchester.
29. Amir, M., Mishra, V., Sharma, R., Iqbal, F., Ali, S. W., Kumar, S., & Khan, G. S. (2023). Development of magnetic nanoparticle based nanoabrasives for magnetorheological finishing process and all their variants. *Ceramics International*, 49(4), 6254–6261.
30. Singh, M., Vajpayee, M., & Ledwani, L. (2020). Eco-friendly surface modification and nanofinishing of textile polymers to enhance functionalisation. *Nanotechnology for Energy and Environmental Engineering*, 529–559.
31. Sankar, M. R., Ramkumar, J., & Jain, V. K. (2009). Experimental investigation and mechanism of material removal in nano finishing of MMCs using abrasive flow finishing (AFF) process. *Wear*, 266(7–8), 688–698.
32. Kathiresan, S., & Mohan, B. (2020). Multi-objective optimization of magneto rheological abrasive flow nano finishing process on AISI stainless steel 316L. *Journal of Nano Research*, 63, 98–111. Trans Tech Publications Ltd.
33. Khan, D. A., Alam, Z., Iqbal, F., & Jha, S. (2019). Experimental investigations on the effect of relative particle sizes of abrasive and iron powder in polishing fluid composition for ball end MR finishing of copper. *International Journal of Precision Technology*, 8(2–4), 354–364.
34. Alam, Z., Khan, D. A., Iqbal, F., & Jha, S. (2019). Effect of polishing fluid composition on forces in ball end magnetorheological finishing process. *International Journal of Precision Technology*, 8(2–4), 365–378.
35. Osama, M., Iqbal, F., Khan, D. A., & Alam, Z. (2022). Design and development of novel multipoint epicyclic superfinishing tool. In *Proceedings of the International Conference on Industrial and Manufacturing Systems (CIMS-2020) Optimization in Industrial and Manufacturing Systems and Applications* (pp. 601–620). Springer International Publishing.

36. Singh, A. K., & Sharma, V. (2021, November). Comparative life cycle assessment of various grinding strategies for Nickel Base superalloys. In *ASME International Mechanical Engineering Congress and Exposition* (Vol. 85567, p. V02BT02A063). American Society of Mechanical Engineers.
37. Alam, Z., Iqbal, F., Ganesan, S., & Jha, S. (2019). Nanofinishing of 3D surfaces by automated five-axis CNC ball end magnetorheological finishing machine using customized controller. *The International Journal of Advanced Manufacturing Technology*, 100, 1031–1042.
38. Iqbal, F., & Jha, S. (2018). Closed loop ball end magnetorheological finishing using in-situ roughness metrology. *Experimental Techniques*, 42, 659–669.
39. Iqbal, F., Alam, Z., Khan, D. A., & Jha, S. (2022). Automated insular surface finishing by ball end magnetorheological finishing process. *Materials and Manufacturing Processes*, 37(4), 437–447.
40. Iqbal, F., Alam, Z., Khan, D. A., & Jha, S. (2019). Part program-based process control of ball-end magnetorheological finishing. In *Advances in Unconventional Machining and Composites: Proceedings of AIMTDR 2018* (pp. 503–514). Singapore: Springer Singapore.

2 Multi-objective optimization of rotational magnetorheological abrasive flow finishing process

Pushpendra Gupta, Vidyapati Kumar, Dilip Kumar Pratihar, and Kalyanmoy Deb

2.1 INTRODUCTION

Nano-finishing, a cutting-edge surface refinement technique, meticulously eliminates a minute fraction of material from the surface, operating within the nano-metre range. This process produces a high-quality surface finish (SF) with excellent surface properties, particularly useful in the electronic and mechanical industries [1, 2]. Nano-finishing methods have various applications in multiple sectors, including aviation, automotive, biomedical, and electronics. In the realm of aviation engineering, impeccably polished copper mirrors find significant utility owing to their exceptional thermal conductivity and heat capacity. However, due to the inherent softness of copper, achieving nano-finishing on this material is highly difficult.

To attain a better SF, researchers applied different magnetorheological (MaR) procedures [3]. Nano-finishing techniques are harnessed to enhance the SF of engine components, yielding substantial benefits such as enhanced fuel efficiency and reduced emissions. Nanotechnology has also been used in paints to protect automobiles from scratches and other damage over time. Nanotechnology is also applied to other body components, including but not limited to chassis, engines, tyres, and drive trains [4]. Within the biomedical industry, nano-finishing plays a pivotal role in augmenting medical implants' SF, fostering improved biocompatibility, and significantly reducing rejection rates. It is required for the mating surfaces of biomedical implants such as shoulder, hip, knee, and ankle joints for appropriate functioning and longevity [5]. In the electronics industry, applying nano-finishing techniques is instrumental in enhancing electronic components' SF, culminating in notable improvements in reliability and overall performance. Nanotechnology has been applied to electronic

DOI: 10.1201/9781003496298-2

components to enhance their performance and durability. Nanotechnology can improve the conductivity of materials used in electronic components, making them more efficient. Nanotechnology can also be used to create smaller and more powerful electronic devices. Figure 2.1 shows the different nano-finishing applications in various industries.

A plethora of nano-finishing techniques exist, encompassing chemical mechanical polishing, electrochemical polishing, electropolishing, magnetic field-assisted finishing, MaR finishing, magnetorheological abrasive flow finishing (MRAFF) [6, 7], and the rotational magnetorheological abrasive flow finishing (R-MRAFF) process [8]. R-MRAFF represents a highly advanced method that harmonizes rotational motion with MaR abrasive flow machining, delivering meticulous SF and precise material removal rates. By employing an MaR fluid infused with abrasive particles and a rotating magnetic field, R-MRAFF effectively eradicates material from the workpiece's surface. Notably surpassing its predecessors, R-MRAFF grants superior control over the material removal rate and SF by harnessing the synergistic potential of rotational motion and MaR abrasive particles.

The capability of generating superior SFs on diverse materials, including hard and brittle substances, has propelled the recent surge of interest in this process. Nonetheless, the optimization of process parameters for achieving the desired output responses in R-MRAFF remains a persistent challenge. To address this, the present study employs elitist non-dominated sorting genetic algorithm II (NSGA-II) [9], a well-established multi-objective optimization algorithm, in an endeavour to optimize the process parameters of R-MRAFF. A detailed discussion about the working of NSGA-II is given in Section 2.4.1. The considered process parameters are

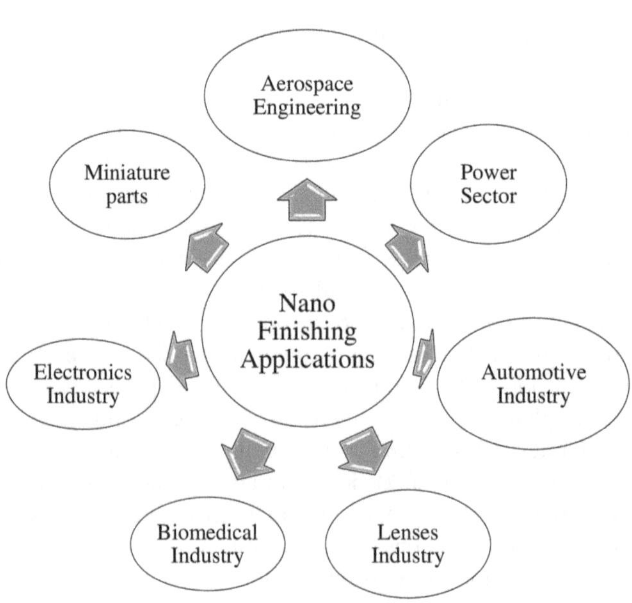

FIGURE 2.1 Applications of nano-finishing process in different industries.

extrusion pressure (*P*) in bars, the number of finishing cycles (*N*), the rotational speed of the magnet (*S*) in RPM, and the mesh size of the abrasive (*M*). Meanwhile, the output responses under scrutiny are the final SF and material removal (MR). Regression analysis is employed to model their relationship to establish a connection between the process parameters and output responses. Initially, a real-coded genetic algorithm maximizes one response by keeping the other as a constraint function. Later, both the responses are maximized using the NSGA-II algorithm and the obtained Pareto-front (PF) shows a clear trade-off between the SF and MR. The proposed approach offers a significant advantage over traditional trial-and-error methods, as it significantly reduces the time and cost associated with optimizing the process parameters of R-MRAFF. Moreover, this approach can be easily extended to other complex manufacturing processes that involve multiple process parameters and output responses. Overall, this chapter aims to provide a comprehensive understanding of the R-MRAFF process and the use of an evolutionary algorithm to optimize its process parameters.

2.2 ROTATIONAL MAGNETORHEOLOGICAL ABRASIVE FLOW MACHINING PROCESS

The preceding section underscored the paramount importance of nano-finishing processes in attaining exceptional surface quality across various sectors. Numerous products' functional requirements necessitate achieving a nano-level SF. Nano-finishing processes can be broadly classified into magnetic field-assisted and non-assisted methods. Magnetic field-assisted techniques encompass magnetic abrasive finishing, MaR finishing, and similar methodologies, while non-assisted methods include abrasive flow finishing. Among the magnetic field-assisted methods, MRAFF stands out as a well-established technique capable of achieving nano-metre SFs on diverse materials, such as brass, aluminium, stainless steel, and silicon nitride. One modification that aids in attaining the desired nano-finishing level involves incorporating rotational motion into the MRAFF process. Figure 2.2(a) illustrates the schematic view of an MRAFF experimental setup. Figure 2.2(b) showcases the R-MRAFF process's mechanism, highlighting the magnets' rotational movement. Additionally, Figure 2.2(c) depicts the alteration in rheological behaviour, specifically the changes in yield stress and viscosity, of the MaR-polishing fluid during the finishing process.

The quality of the finished surface in MaR-fluid-based finishing processes is primarily influenced by the constituents of the MaR fluid and the applied magnetic field [11]. In their work, Sidpara and Jain [11] explored the correlation between rheological properties and SF quality in MaR-fluid-based finishing processes. R-MRAFF employs an MaR fluid that possesses a distinctive characteristic of altering its viscosity and yield strength when it is subjected to a magnetic field. The magnetic field induces the formation of a microstructure within the fluid, enabling manipulation of the flow rate and direction. The presence of abrasive particles within the fluid facilitates the removal of the material from the workpiece's surface. The concurrent rotation of the magnetic field ensures an even distribution of the abrasive particles, resulting in a uniform SF. Consequently, the process utilizes an MaR fluid infused

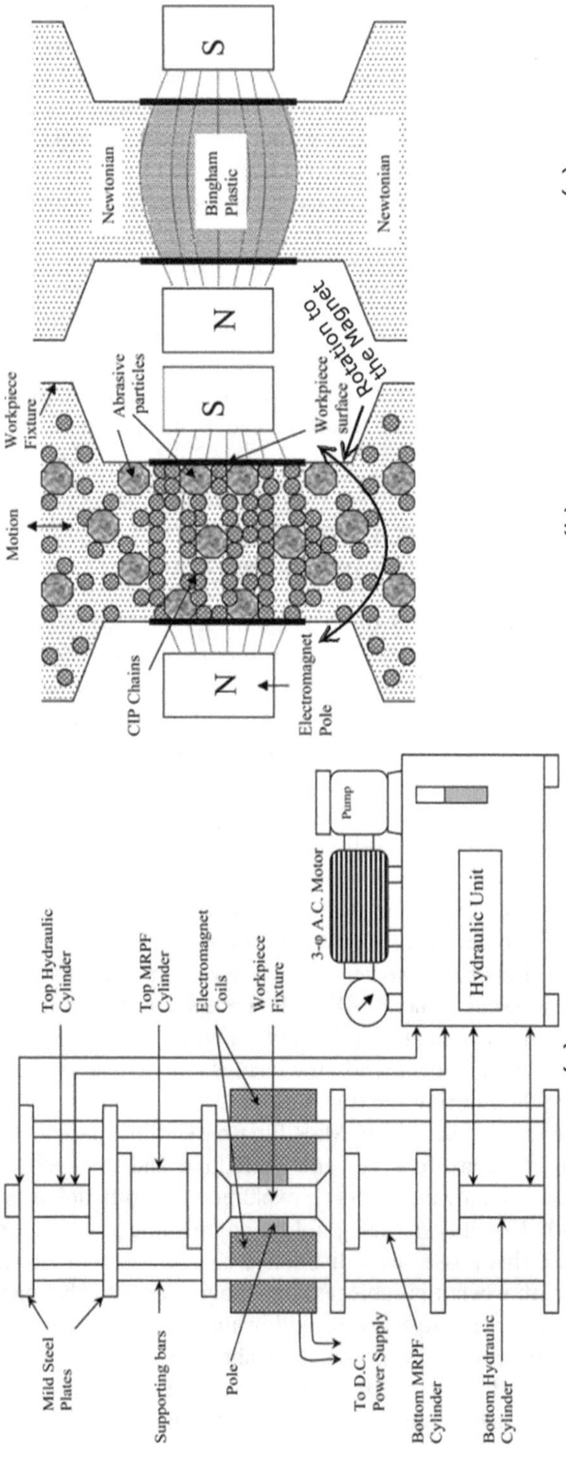

FIGURE 2.2 (a) A schematic representation of the experimental setup for MRAFF; (b) mechanism of the R-MRAFF process; (c) observed changes in the rheological behaviour of the MaR-polishing fluid during the finishing process [10].

with abrasive particles and a rotating magnetic field to effectively eliminate material from the workpiece's surface. The distinctive amalgamation of rotational motion and MaR abrasive particles enables precise control over the material removal rate and SF, rendering it an ideal process for nano-finishing applications. R-MRAFF emerges as a significant nano-finishing technique, given its unparalleled ability to achieve high-quality SFs while ensuring precise control over the material removal rate. Compared to earlier processes, R-MRAFF offers enhanced control over the MR rate and SF, making it an appealing choice for accomplishing exceptional nano-level finishes.

2.3 RECENT DEVELOPMENTS IN THE MRAFF PROCESS

Rajput et al. [12] conducted a thorough examination of the different procedures and techniques used in MaR-fluid finishing processes and an assessment of their efficacy too. The findings showed that this technique may produce extraordinary accuracy and surface quality without affecting the product's material topography. Karthikeyan et al. [13] investigated the in vitro behaviour of stainless steel 316 L treated to R-MRAFF in their investigation. They evaluated surface topography, roughness (Ra), and microhardness using experiments and analyses of treated materials. Their results showed that the procedure improved surface quality, decreased roughness, and maintained or increased microhardness. These findings imply that the R-MRAFF technique has the potential to improve the performance of stainless-steel components in a variety of applications. In addition, Rajput et al. [14] devised a feature-based hybrid MRF planning system (FHMRF-PS) for the automated process planning of paraffin wax deposition in workpiece holes and pocket features prior to MRF. A case study on a Ti-6Al-4 V bone plate was performed to assess the efficiency and efficacy of the FHMRF-PS. The proposed process planning during MRF reduced the initial Ra value from 324.12 nm to 21.56 nm, illustrating the effectiveness of the designed FHMRF-PS.

Kumar et al. [15] discovered that an F-to-S iron ratio of 1:1, a DC supply of 12 V, a work sample rotating speed of 200 rpm, and a linear feed rate of 0.1 mm/min were the best parameter combinations for producing high-quality surfaces on S.S. 316 L cylindrical work samples. Karthikeyan et al. [16] elevated the hemocompatibility of SS 316 L by lowering surface roughness and boosting surface energy during the R-MRAFF process. The required surface polish of SS 316 L was attained using the following process parameters: 0.5-tesla magnetic field intensity, 10% abrasive concentration, and 200 revolutions per minute of rotating speed. Choopani et al. [17] encountered the effectiveness of the R-MRAFF technique in obtaining the requisite nano-finishing for Al2024 tubes. Surprisingly, they attained an Ra value of 26.3 nm and an MR value of 41 mg in just 15 minutes, demonstrating the efficiency of this method. Furthermore, Kumar et al. [18] implemented the R-MRAFF technique to obtain a minimum Ra of 34.5 nm and a maximum percentage increase in Ra ($\%\Delta R_a$) of 85.56% on a small steel gear. Furthermore, the finishing procedure efficiently removed the production flaws.

In light of the reviewed research, it becomes evident that MaR-fluid finishing techniques possess significant potential. These techniques demonstrate the ability to achieve exceptional precision, enhance surface quality, reduce roughness, and improve material performance across diverse applications. The optimization of process parameters emerges as a critical factor, enabling the attainment of desired SFs and the elimination of manufacturing flaws. Thus, it is crucial to gain a comprehensive understanding of and effectively control the variables involved in these processes to ensure favou outcomes.

2.4 ELITIST NON-DOMINATED SORTING GENETIC ALGORITHM II

Elitist NSGA-II represents a state-of-the-art optimization algorithm rooted in evolutionary computing. Widely regarded as a benchmark technique in evolutionary computation, NSGA-II operates as a multi-objective optimization algorithm, enabling the simultaneous optimization of multiple objectives. Initially proposed by Deb et al. [9] in 2002 as an enhancement to the original NSGA algorithm, NSGA-II stands as a fast, efficient, and robust algorithm that has found extensive application in engineering, science, and economics. Its high efficiency and robustness make it an indispensable tool for optimizing complex systems and processes in industrial settings, capable of handling diverse optimization problems, including those involving non-linear and non-convex objective functions.

Unlike single-objective optimization algorithms that aim to discover a single optimal combination of design variables, a multi-objective optimization algorithm, such as NSGA-II, seeks to uncover multiple solutions known as Pareto-optimal solutions. These solutions form a set of "non-inferior" solutions in the objective space, wherein further improvement in one objective necessitates a compromise in at least one of the other objectives. Addressing the needs of most industrial applications, NSGA-II tackles the simultaneous resolution of two or more conflicting objectives, generating better Pareto-optimal solutions. By minimizing two or three objectives concurrently and accommodating constraints, NSGA-II produces a Pareto-optimal set with numerous solutions that optimize both objectives to such an extent that no other solution can enhance any objective without deteriorating another.

The primary advantage of NSGA-II lies in its capacity to identify a set of solutions that are non-dominated within the search space. Consequently, the algorithm discovers multiple solutions that are optimal in distinct ways rather than just a single optimal solution. This characteristic proves especially valuable in multi-objective optimization problems where the simultaneous optimization of multiple objectives is necessary, and no single solution can optimize all of them. The algorithm commences by generating an initial population of solutions evaluated based on their fitness in the search space. The fitness of a solution is determined by its ability to satisfy the various objectives of the optimization problem. Subsequently, the solutions are sorted into different levels of non-domination, with the first level consisting of solutions that are not dominated by any other solution, followed by subsequent levels containing solutions that are dominated by those in the preceding level, and so forth. After sorting the solutions, NSGA-II employs selection, crossover, and mutation operators to generate a new population of solutions that remain non-dominated within the search space. This iterative process continues until optimal solutions are discovered or the maximum number of iterations is reached.

2.4.1 Working of NSGA-II

NSGA-II operates by maintaining a population of candidate solutions and iteratively guiding them towards improved outcomes, making it an exemplary algorithm for multi-objective optimization. One of its key strengths lies in its ability to converge towards the true PF, which is accomplished through the effective utilization of crossover and mutation operators. Crossover combines two chromosomes to create offspring, while mutation introduces random changes at the gene level. This interplay between crossover and mutation enables the population to converge while simultaneously reintroducing genetic diversity to steer clear of local minima. In addition to convergence, NSGA-II achieves other important qualities, such as maintaining a uniformly distributed and diverse set of solutions. These objectives are facilitated by the implementation of crowding distance. By employing an efficient crowding-distance assignment approach, NSGA-II ensures a well-spread population of solutions, while the employment of fast, non-dominated sorting enhances the algorithm's convergence capabilities. These advantages have propelled NSGA-II into wide-ranging applications and subjected it to numerous comparative studies, validating its performance. NSGA-II favours solutions with a higher crowding distance over other solutions within the same PF. It also incorporates the concept of elitism, where the best solutions are safeguarded and carried over to the next generation. These elite solutions are never eliminated and directly advance to the subsequent generation. The remaining solutions are compared amongst themselves, considering both their crowding distance and their rank, to determine the most favourable solution. This ensures a fixed population size, as illustrated in Figure 2.3. Here, P_t is the set of solutions obtained in generation t and Q_t is the solution set obtained using crossover and mutation operators on P_t. Non-dominated sorting is performed after considering offspring and parent solutions to get different PFs. Crowding distance is used to maintain the diversity of the solutions on each front.

To obtain the PF for multi-objective optimization, NSGA-II follows a step-by-step procedure:

1. Generate an initial random population, P.
2. Evaluate each solution to obtain the fitness function.
3. Perform non-dominated sorting based on non-domination criteria to find different PFs.
4. Calculate crowding distance to preserve diversity on each front.
5. Select individuals for the next generation based on their rank (lowest) and crowding distance (highest).
6. Generate new solutions, Q, from the previous generation using two operators, namely, crossover and mutation.

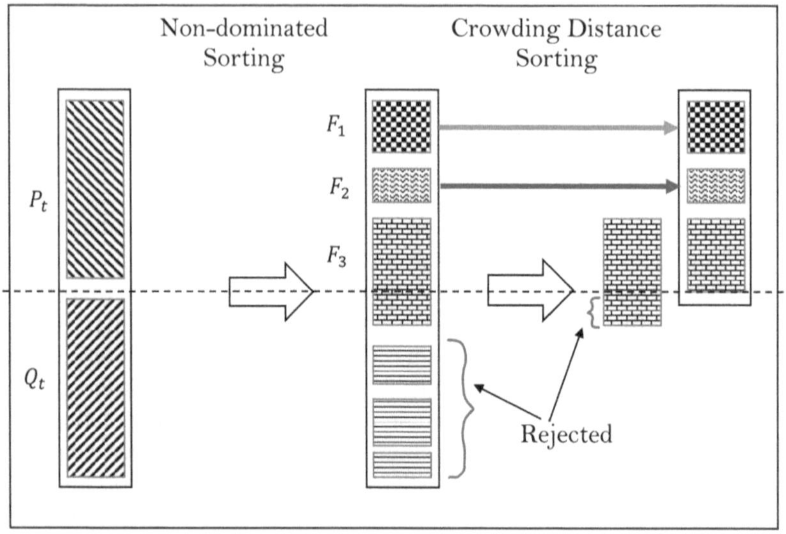

FIGURE 2.3 Working of NSGA-II.

The NSGA-II algorithm repeats steps 2 through 6 until a predefined stopping criterion is met. Typically, this criterion is either reaching the maximum number of generations or failing to obtain a better PF compared to the previous iteration.

2.5 PRESENT STUDY

In various industrial sectors, the design of experiments (DOE) [19–23] is frequently employed for optimizing and developing manufacturing processes by examining the effect of multiple factors on a response. Das et al. [8] employed a DOE to evaluate the summary of responses, specifically MR and SF, using five levels of process parameters in their study on the R-MRAFF process. The R-MRAFF method was utilized to finish the internal surface of cylindrical stainless-steel (non-magnetic) workpieces. The central composite design (CCD) employed four factors with five levels, yielding a total of 30 experiments comprising 16 factorial runs, 8 axial runs, and 6 centre runs. The first parameter, hydraulic extrusion pressure (*P*) in bars, was varied across five levels, namely, 32.5, 35, 37.5, 40, and 42.5. The second parameter, the number of finishing cycles (*N*), ranged from 600, 800, 1000, 1200 to 1400. The third parameter, the rotational speed of the magnet (*S*) in RPM, varied across 50, 100, 150, 200, and 250. The fourth parameter, the mesh size of the abrasive (*M*), ranged from 90, 120, 150, 180 to 210. Table 2.1 displays the responses along with their various levels of process parameters considered during the experimental design.

Table 2.1 includes measurements for both initial and final SF, and the percentage improvement in roughness denoted by $\%\Delta R_a$ as shown in the sixth column. Additionally, the material removal (MR) was calculated by weighing the samples before and after the machining process, with the difference between the two measurements indicating the amount of material removed in mg from the sample.

TABLE 2.1
Experimental results of R-MRAFF process on surface finish and material removal [8]

Run no.	Extrusion pressure, *P* (bar)	Finishing cycle, *N*	RPM magnet, *S*	Mesh size of SiC, *M*	Surface finish (in %ΔR_a)	Material removal (mg)
1	35	1200	100	120	90.00	12.3
2	40	1200	200	180	90.48	13.7
3	35	1200	200	180	81.08	8.1
4	37.5	1000	150	150	91.30	20.9
5	40	1200	200	120	91.89	23.5
6	40	800	200	180	74.29	7.3
7	40	800	100	180	87.23	11.4
8	37.5	600	150	150	81.08	6.8
9	37.5	1000	150	210	82.05	5.0
10	37.5	1000	150	150	90.91	25.1
11	37.5	1400	150	150	85.71	26.3
12	37.5	1000	150	150	90.00	23.2
13	42.5	1000	150	150	90.70	15.3
14	40	800	100	120	86.21	4.1
15	35	1200	200	120	80.00	24.0
16	37.5	1000	250	150	76.74	28.2
17	35	800	200	120	71.43	20.5
18	32.5	1000	150	150	75.86	8.1
19	40	1200	100	180	95.15	20.5
20	40	800	200	120	81.25	20.3
21	37.5	1000	50	150	82.22	5.1
22	37.5	1000	150	150	89.74	24.5
23	35	800	100	120	93.94	8.62
24	37.5	1000	150	150	85	20.5
25	37.5	1000	150	150	92.31	18.5
26	35	1200	100	180	92.17	19.9
27	40	1200	100	120	92.31	18
28	35	800	200	180	73.91	7.4
29	37.5	1000	150	90	88.24	27.8
30	35	800	100	180	85.29	11.5

2.6 RESULTS AND DISCUSSION

Table 2.1 shows the SF and MR values obtained from the R-MRAFF experiments, which involved four factors at five levels using the CCD design. Figure 2.4 shows how the response variations are affected by each input parameter, using the mean values of both responses for a given process parameter level. During the increase in hydraulic pressure (*P*) from 32.5 to 35 bar and 35 to 37.5 bar, MR increased by 73.3% and 37.6%, respectively. Meanwhile, SF increased by 10.0% and 3.3%, respectively.

However, further increasing *P* from 37.5 bar to 40 bar caused a decrease in MR by 23.2% and an increase in SF by 1.2%. Increasing *P* from 40 to 42.5 bar increased MR and SF by 3.0% and 3.8%, respectively. While increasing the number of finishing cycles (*N*) from 600 to 800 and 800 to 1000, MR increased by 67.5% and 62.56%, respectively, while SF increased by 0.75% and 5.5%. Increasing *N* from 1000 to 1200 caused a decrease in MR by 5.5% and an increase in SF by 3.3%.

Increasing *N* from 1200 to 1400 resulted in an increase in MR by 50.28% and a decrease in SF by 3.8%. When the magnet's rotational speed (*S*) is increased from 50 to 100 RPM, the MR increased significantly by 160.6%, while the improvement in SF was only 9.8%. As *S* was raised from 100 to 150 RPM, MR increased by 39.2%, but SF decreased by 3.7%. However, when *S* increased again from 150 to 200 RPM, MR and SF worsened, declining by 15.7% and 7.32%, respectively. Conversely, when *S* is increased from 200 to 250 RPM, MR increased by 80.8%, but SF decreased by 4.7%. The increase in the mesh size of the abrasive (*M*) from 90 to 120 has an adverse impact on both MR and SF, leading to a reduction of 41% and 2.7%, respectively. However, an increase in *M* from 120 to 150 resulted in an MR increase of 13%, while it barely affected the SF (almost 0.1%). Subsequently, an increase in *M* from 150 to 210 led to a decline in both MR and SF. The MR is decreased by 32% and 60%, respectively, while the SF is reduced by 1.2% and 3.4%, respectively.

This analysis of experimental data examined the effects of the process parameters on the response variation, and it has been discovered that MR is more strongly affected than SF due to changes in process parameters. This is so because MR measured the total amount of material removed (in mg) from the surface, while SF focused on SF at a nano level, which has minimal variation. The least effect on MR is observed due to parameter *P*, whereas the least impact on SF is observed due to

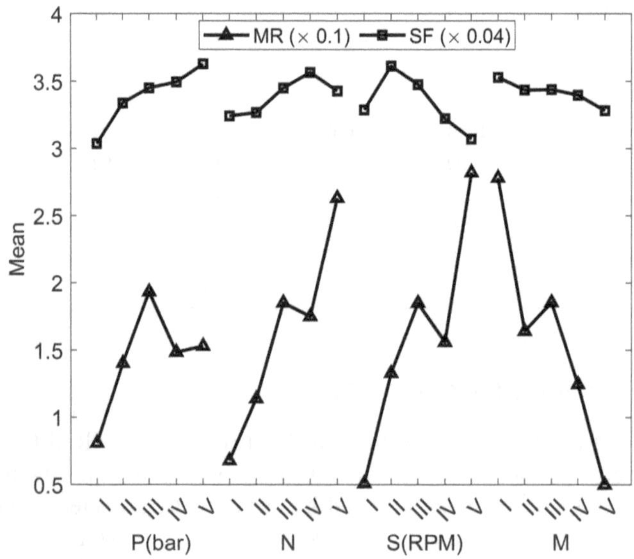

FIGURE 2.4 Effect of process parameters on response variation.

parameter M. Parameter S has a significant effect on both MR and SF at different levels. If we consider the changes in responses caused by altering the process parameters to different levels, regardless of whether the responses increased or decreased, the order in which the parameters impacted MR from the least to the most is M, N, P, and S, while for SF it is M, N, P, and S.

2.6.1 Mathematical modelling

The regression equation for MR and SF are developed, and the coefficient of determination (R^2) for both are found to be approximately 83.93% and 80.72%, respectively. Using regression analysis, the dependent variables MR and SF are modelled as non-linear functions of four independent process parameters, which are given in Equations (2.1) and (2.2).

$$\begin{aligned}\mathrm{MR} = &-688 + 30.4P + 0.0382N + 0.761S + 0.693M - 0.440\,P^2 - 0.000038\,N^2 - \\ &0.000605S^2 - 0.001751\,M^2 + 0.00204\,PN + 0.00156\,PS + \\ &0.0046\,PM - 0.000133\,NS + 0.000003\,NM - 0.003003\,SM\end{aligned} \tag{2.1}$$

$$\begin{aligned}\mathrm{SF} = &-54 + 10.65P - 0.0804N - 0.560S + 0.111M - 0.195P^2 - 0.000030N^2 - \\ &0.000867\,S^2 - 0.000835\,M^2 + 0.00277PN + 0.01599\,PS - \\ &0.0013\,PM + 0.000160\,NS + 0.000175\,NM - 0.000091\,SM\end{aligned} \tag{2.2}$$

The comparison between the present regression analysis and Das et al.'s [8] model is also shown in Figure 2.5. The MR equations are closely related and well-fitted to all experimental data. However, there is a significant difference in the SF equations, with Equation (2.2) fitting all experimental data closely, while the previous model barely passed through any data point. The present model has clearly outperformed the earlier model in terms of goodness of fit.

2.6.2 Effect of process parameters on responses

The variation of MR and SF with respect to the input parameters is shown in Figures 2.6 and 2.7, as response surface plots. The variation is shown concerning two independent values, while the other two are kept constant. Figure 2.6(a) shows that MR initially increases with increasing pressure (P) and finishing cycles (N). However, MR eventually decreases with further increases in P and N. The maximum MR is found to occur when P and N are approximately equal to 37.5 bar and 1000, respectively. The other two parameters, the magnet's rotational speed (S) in RPM and mesh size (M), are kept fixed. Figure 2.6(b) shows that MR increases with increasing S up to 200 RPM. After that, MR slightly decreases. N causes MR to

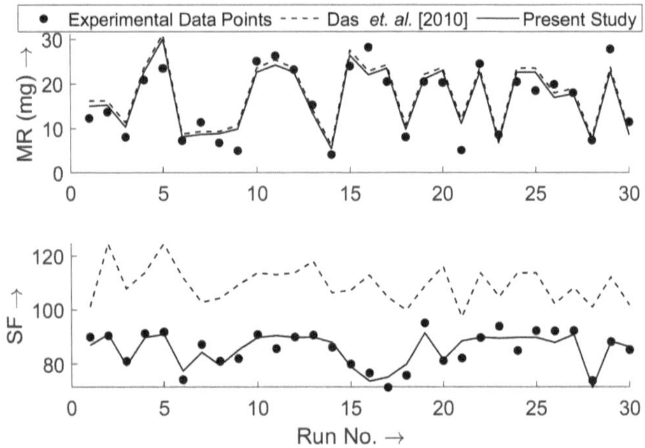

FIGURE 2.5 Comparison of the goodness of fit of the regression equations on the experimental data points between the present study and the study conducted by Das et al. [8].

increase exponentially at a fixed *S* (at 50 RPM). However, increasing *S* and *N* together causes MR to reduce further. Figure 2.6(c) shows that at a fixed *M* (around 100), the parameter *S* is found to be increasing the MR exponentially.

Similarly, at a fixed value to *S* (around 50 RPM), *M* is also found to be increasing the MR. However, the effect of *S* is found to be more compared to *M*. In the case of increasing both the parameters *M* and *S*, MR is found to be decreased drastically; in fact, it reached a minimum. Figure 2.6(d) shows that MR achieved an optimum value when *P* reached around 37.5 bar and *M* at around 160. Figure 2.6(e) shows a similar trend with respect to *P*. However, parameter *S* is found to be significantly helping in achieving the better MR with the higher value of *S*. Here, *P* is also found to assist in achieving the better MR at around 37.5 bar, and at this value of *P*, the parameter *S* further increases the MR with an advance in its own value. Figure 2.6(f) shows a very similar trend with the parameters *N* and *M*, which was observed in Figure 2.6(e) with the parameters *S* and *P*, respectively. Here, the best value for *M* is found to be around 150, and *N* increases the MR by increasing its own value.

Similarly, Figure 2.7 shows the response surface plot for finish (SF). Figure 2.7(a) shows that the parameters *P* and *N* positively affect achieving the better SF. Figure 2.7(b) shows that the parameter *S* initially helps to attain the better SF but reduces it drastically later. At around *S* equals 150 RPM, SF seems to improve with increasing *N*. Figure 2.7(c) shows that the parameter *M* initially improves SF and then reduces it. However, the parameter *S* shows a similar trend, as mentioned in Figure 2.7(b); for example, at around *S* = 150 RPM, it enhances SF. Figure 2.7(d) shows that when *P* is kept around 37.5 bar, SF seems to improve with increasing *M*, provided that the rest of the parameters are kept fixed. Figure 2.7(e) shows that increasing *S* and *P* could help to obtain better SF, while *M* and *N* are kept fixed. Figure 2.7(f) shows a similar trend with the parameters *M* and *N*. The results show that both MR and SF are complex functions of the input parameters.

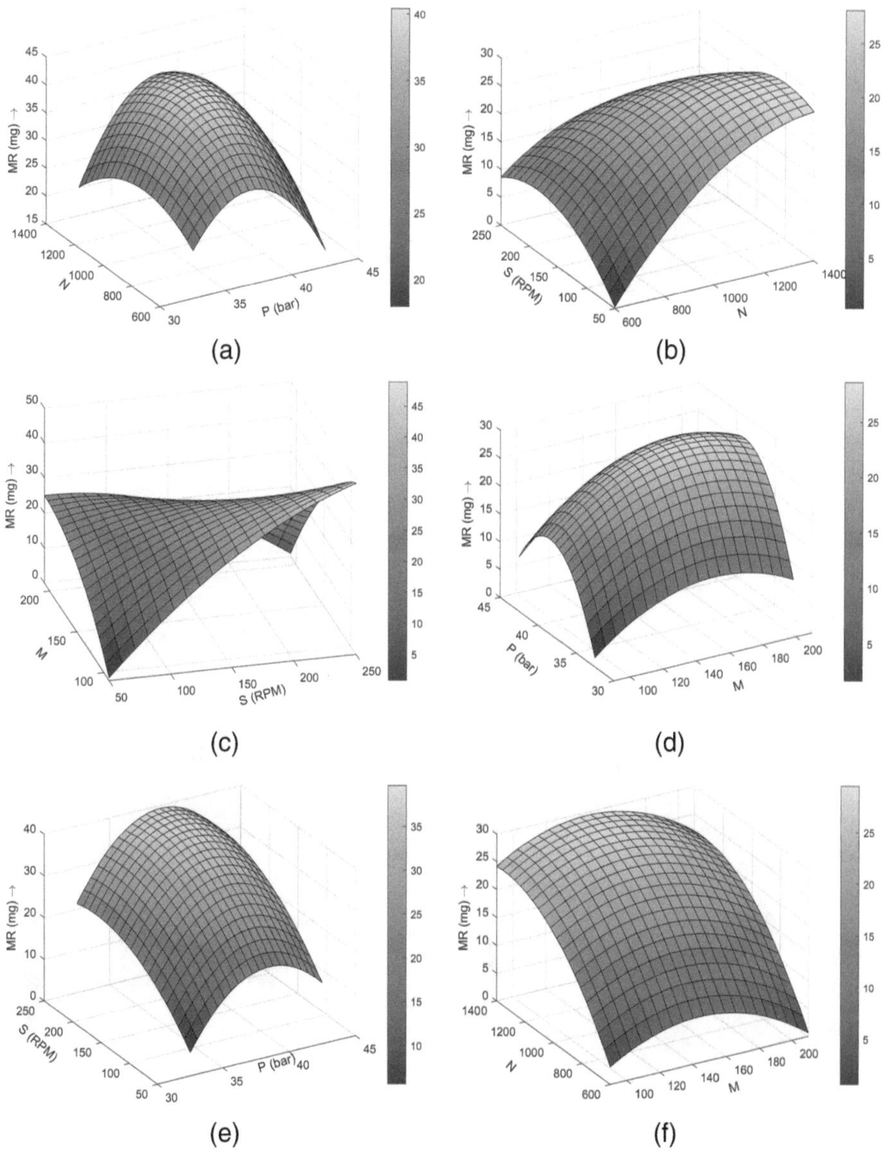

FIGURE 2.6 Effects of variation of process parameters on material removal.

2.6.3 Single-objective optimization

The results of single-objective optimization using a real-coded genetic algorithm [24] with a probability of crossover and mutation set to 0.9 and 0.3, respectively, are shown in Table 2.2. The population size and number of generations used are 200 and 500, respectively. Initially, SF is constrained to be at least 80%, while MR is maximized. The optimal values of the decision variables are found to be as follows: $P = 39.14$, $N = 1175$,

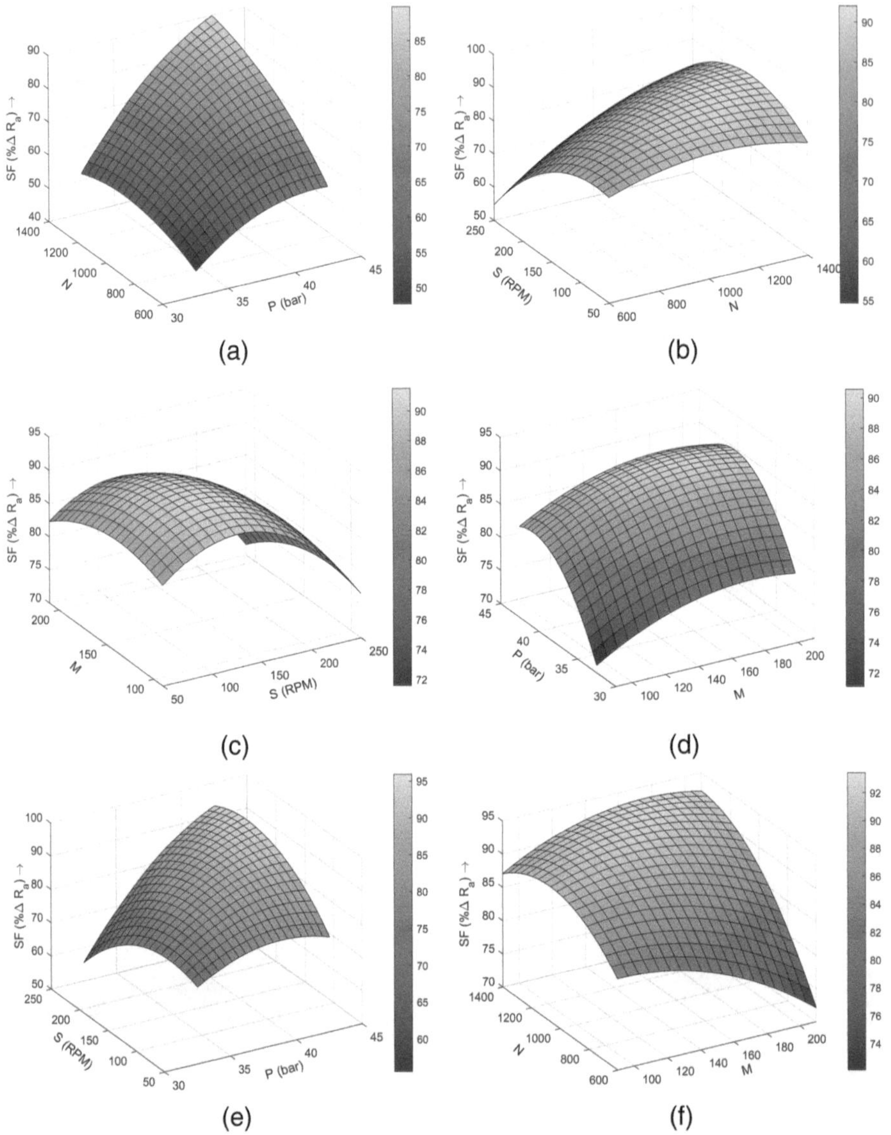

FIGURE 2.7 Effects of variation of process parameters on SF.

$S = 250$, and $M = 90$. In the second attempt, SF is maximized while keeping MR as a constraint function. The constraint required a minimum material removal of 24 mg from the samples. The best combination of process parameters to achieve a 95.87% improvement in roughness is found to be as follows: $P = 41.34$, $N = 1395$, $S = 182$, and $M = 137$. This clearly demonstrates the trade-off between the two objective functions.

However, these two solution sets represent the only options for prioritizing one objective function by keeping the other as a constraint. To visualize the trade-off

TABLE 2.2
Single-objective optimization results using a genetic algorithm (GA)

Objective functions (maximized)	Optimal design variables				Objective functions	
	P (bar)	*N*	*S* (RPM)	*M*	MR (mg)	SF (%ΔRa)
MR (mg)	39.14	1175	250	90	39.81	80.00
SF (%ΔR_a)	41.34	1395	182	137	24.00	95.87

better, the NSGA-II algorithm is used to identify a set of Pareto-optimal solutions that simultaneously maximize both MR and SF. NSGA-II is a population-based algorithm that searches for Pareto-optimal solutions iteratively.

2.6.4 Multi-objective optimization

In nano-finishing operations, the goal is to achieve high levels of both SF and material removal (MR). However, these objectives can often conflict with each other. Increasing MR may result in a decrease in SF and vice versa. As a result, it is not possible to maximize both objectives simultaneously. To address this trade-off, the NSGA-II algorithm is used to simultaneously maximize both MR and SF. The NSGA-II algorithm is designed to minimize the objectives by default. Therefore, the duality principle is used to convert the problem into a minimization problem by taking the reciprocal of MR and SF. The objectives are expressed using a quadratic mathematical model with four decision variables as given in Eqs. (2.1) and (2.2). The NSGA-II multi-objective problem is formulated as follows:

$$\text{Minimize } \frac{1}{\text{MR}} \text{ (refer to Eq. (2.1))}$$

$$\text{Minimize } \frac{1}{\text{SF}} \text{ (refer to Eq. (2.2))}$$

with decision variables

$$P, N, S, \text{ and } M$$

subject to

$$32.5 \leq P \leq 42.5 \text{ (bar)},$$

$$600 \leq N \leq 1400,$$

$$50 \leq S \leq 250 \text{ (rpm)},$$

$$90 \leq M \leq 210$$

Out of the four decision variables, variation of pressure (*P*) is considered a real decision variable, while the rest are considered as integer decision variables. The

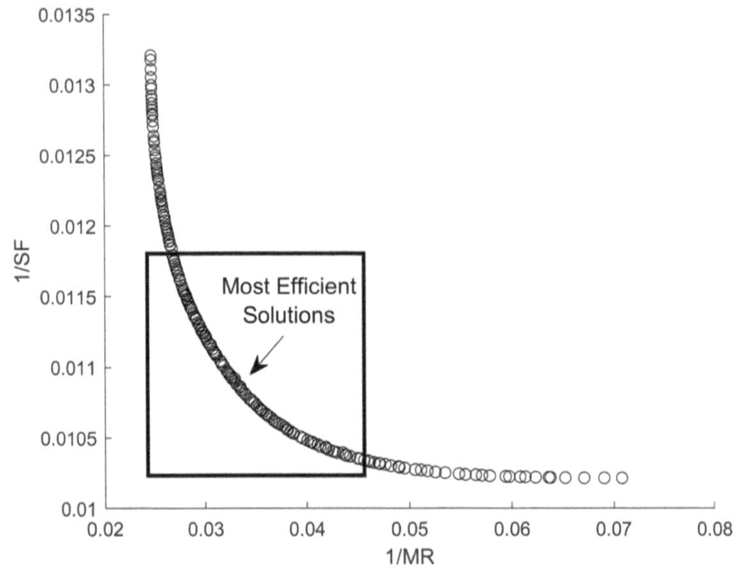

FIGURE 2.8 NSGA-II Pareto-optimal solution set in objective space.

multi-objective optimization problem described above is solved using NSGA-II with the following parameters:

Population Size = 200,
Number of Generations = 500,
Probability of Crossover = 0.9,
Probability of Mutation = 0.25,
Distribution Index for Crossover = 10, and
Distribution Index for Mutation = 20.

The Pareto-optimal solutions generated by NSGA-II are presented in Figure 2.8, which can be used to select the optimal machining parameters based on the decision maker's requirements. The most efficient solutions are enclosed within a rectangular box in Figure 2.8, as beyond that region in either direction, there is a higher sacrifice of one objective to achieve a marginal improvement in the other.

The obtained PF, as shown in Figure 2.8, is found by the variation of decision parameters, which are shown in Figure 2.9. The graph's x-, y-, and z-axes represent the variation of parameters N, P, and S, respectively. The parameter M has been divided into four groups, namely, 90–120, 121–150, 151–180, and 181–210. However, none of the solutions from the 181–210 group are identified, marked by an asterisk (*) in Figure 2.9.

It is observed that increasing the mesh size has a negative impact on both MR and SF. The relationship between the non-dominated solutions obtained using different decision variables can be better understood by examining Figure 2.9. It is evident that only a specific range of values for the parameter P (between 37 and 42.5 bar)

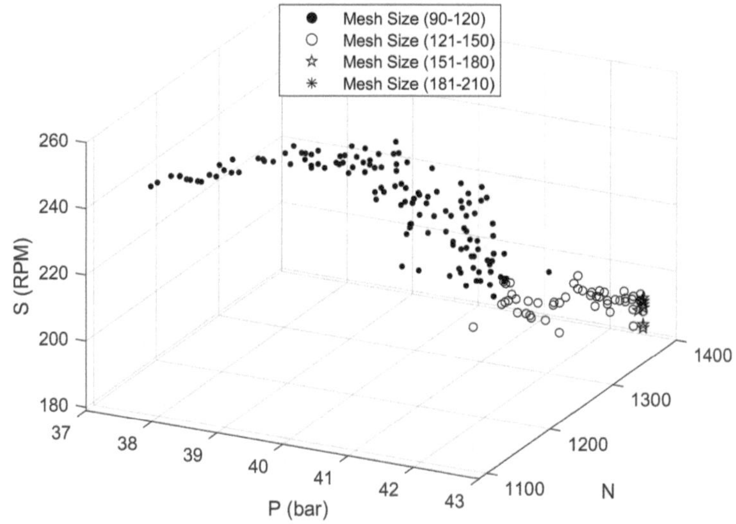

FIGURE 2.9 Variation of pressure (*P*), finishing cycles (*N*), magnet's rotation speed (*S*), and mesh size (*M*) in decision space.

significantly impacts both the objectives, whereas the initial two levels of *P* are not effective for either objective. The last two levels of *N* and *S* parameters are highly effective in achieving good SF and MR. These effects are further studied in detail using Figure 2.10.

Figure 2.10 displays the effects of *P*, *N*, *S*, and *M* parameters on MR and SF. Figure 2.10(a) reveals that when *P* is set to 37.5 bar, MR is maximized, but SF is poor. As *P* is increased, MR decreases, and SF improves. NSGA-II generates several solutions that can achieve good SF values close to 93% and MR values close to 32.5 mg at around 40.5 bar. SF was enhanced with an increase in *P*, as the higher axial force facilitated material removal in the form of microchips. MR is increased due to the deeper penetration of abrasive particles into the workpiece surface, as the radial force is increased at the higher *P* [8]. However, MR reached a maximum at an optimum value of 38 bar and then decreased as the depth of penetration reduced due to the shear-thinning nature of the MR polishing fluid. Figure 2.10(b) shows that the value of *N* that is good for both the objectives is approximately equal to 1300. Increasing *N* further reduces MR, but it improves SF. SF is enhanced with an increase in the parameter *N* due to a continuous enhancement in surface texture. However, the abrasive particles that initially cut the peaks of rough surfaces become worn out as *N* increases, and their effects on the polished surface are reduced due to increased hardness. Thus, it resulted in a poor amount of MR.

Figure 2.10 (c) shows that when *S* is kept equal to 210 RPM, MR and SF are both found to be good. However, increasing or decreasing *S* adversely affects both the objective functions. SF attains its maximum value at around 180 RPM of the parameter *S*. Beyond this value, SF is decreased, while MR increases. This trend can be attributed to increased tangential cutting force, which enhances MR. However, the centrifugal force acting on the abrasive particles indents the workpiece surface,

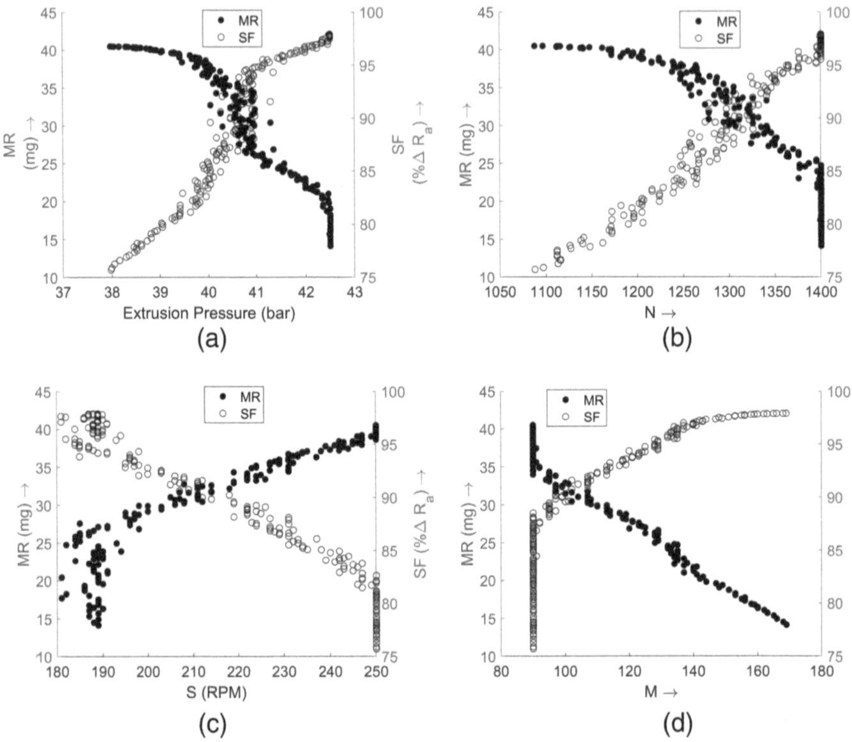

FIGURE 2.10 Effect of pressure (*P*), finishing cycles (*N*), magnet's rotation speed (*S*), and mesh size (*M*) on material removal (MR) and SF.

resulting in a poorer SF [8]. Lastly, as seen in Figure 2.10 (d), several solutions can attain a good SF value as high as 87.5% and MR close to 35 mg, even at a mesh size of 90. As the mesh size increases, only fewer solutions are available, resulting in a better SF but a poorer MR. Both SF and MR improved at lower *M*. The higher *M* leads to more abrasives in the polishing medium due to a fixed concentration, and the abrasive diameter decreases at the larger mesh sizes. Consequently, the indentation force acting on each abrasive particle diminishes with increased mesh size, reducing the MR. However, SF reached its maximum value at approximately 170 RPM.

2.7 CONCLUSION

R-MRAFF continues to emerge as an enticing choice for achieving exceptional nano-level SFs, boasting a uniform, smooth mirror-like surface (SF) with remarkable precision at the nano-metre scale. Moreover, this process exhibits an impressive capacity for higher material removal (MR), surpassing the capabilities of preceding methods. By harnessing the synergistic effects of rotational motion and MaR abrasive particles, R-MRAFF grants superior command over the material removal rate and SF, setting it apart as an advanced solution in the realm of nano-finishing techniques. The significance of R-MRAFF extends beyond its exceptional performance, as it enables the selective finishing of both internal and external surfaces on intricately shaped workpieces. Furthermore, this technique proves invaluable in situations where conventional finishing methods struggle to access and address specific areas. Consequently, its potential spans a wide array of industries that demand meticulous SFs on cylindrical components. Notably, sectors such as aerospace, automotive, and medical industries get significant benefits from the transformative capabilities of R-MRAFF.

In light of the comprehensive study conducted, the following conclusions have been drawn, shedding light on the optimal parameters and their corresponding outcomes:

- The optimal combination of parameters *P*, *N*, *S*, and *M* is found to be 39.14 bar, 1175 cycles, 250 rpm, and 90 mesh size, respectively, to achieve the highest MR of 39.81 mg, while SF is improved by around 80%.
- The optimal combination of process parameters to achieve SF of 95.87% is found when the parameters *P*, *N*, *S*, and *M* are set to 41.34 bar, 1395 cycles, 182 rpm, and 137 mesh size, respectively. The value of MR at the same parameter setting is found to be 24 mg.
- The higher levels of *P*, *N*, and *S* create a trade-off between improving the SF and obtaining a high MR. However, the lower levels of *M* are suitable for both responses. This can help the decision maker to select other combinations of process parameters for the desired result in SF and MR.
- Only a few solutions are able to improve SF at the expense of lower MR, which suggests that enhancing SF is much more difficult than increasing the material removal using different combinations of parameters.

In summary, R-MRAFF emerges as an exceptionally promising approach for achieving superior nano-level SFs, surpassing previous processes in terms of uniformity, precision, and material removal efficiency. Its application potential spans diverse industries, including aerospace, automotive, and medical sectors, where high-quality SFs on cylindrical components are in high demand. Through the implementation of NSGA-II optimization, this study not only highlights the efficacy of R-MRAFF but also demonstrates the paramount importance of optimizing process parameters to attain the desired SF and material removal outcomes. By combining the strengths of R-MRAFF and advanced optimization techniques, this research paves the way for advancements in SF technology and opens new horizons for impeccable surface quality in various industrial applications.

REFERENCES

1. A. K. Singh, S. Jha, and P. M. Pandey, "Mechanism of material removal in ball end magnetorheological finishing process," *Wear*, vol. 302, no. 1–2, pp. 1180–1191, Apr. 2013, doi: 10.1016/J.WEAR.2012.11.082.
2. S. Jha and V. K. Jain, "Modeling and simulation of surface roughness in magnetorheological abrasive flow finishing (MRAFF) process," *Wear*, vol. 261, no. 7–8, pp. 856–866, Oct. 2006, doi: 10.1016/J.WEAR.2006.01.043.

3. H. Kansal, A. K. Singh, and V. Grover, "Magnetorheological nano-finishing of diamagnetic material using permanent magnets tool," *Precis Eng*, vol. 51, pp. 30–39, Jan. 2018, doi: 10.1016/J.PRECISIONENG.2017.07.003.
4. J. Mathew, J. Joy, and S. C. George, "Potential applications of nanotechnology in transportation: A review," *J King Saud Univ Sci*, vol. 31, no. 4, pp. 586–594, Oct. 2019, doi: 10.1016/J.JKSUS.2018.03.015.
5. T. S. Bedi, R. Kant, and H. Gurung, "Advanced finishing processes for biomedical applications," *Materials Horizons: From Nature to Nanomaterials*, pp. 105–126, 2022, doi: 10.1007/978–981-16–3645-5_5.
6. A. Barman and M. Das, "Generation of nano-level surface finish by advanced nanofinishing processes," *Lecture Notes in Mechanical Engineering*, pp. 199–214, 2020, doi: 10.1007/978–981-15–2117-1_10/COVER.
7. M. Kumar, A. Alok, V. Kumar, and M. Das, "Advanced abrasive-based nano-finishing processes: challenges, principles and recent applications," *Materials and Manufacturing Processes*, vol. 37, no. 4, pp. 372–392, 2021, doi: 10.1080/10426914.2021.2001509.
8. M. Das, V. K. Jain, and P. S. Ghoshdastidar, "Nano-finishing of stainless-steel tubes using rotational magnetorheological abrasive flow finishing process," *Machining Science and Technology*, vol. 14, no. 3, pp. 365–389, Jul. 2010, doi: 10.1080/10910344.2010.511865.
9. K. Deb, A. Pratap, S. Agarwal, and T. Meyarivan, "A fast and elitist multiobjective genetic algorithm: NSGA-II," *IEEE Transactions on Evolutionary Computation*, vol. 6, no. 2, pp. 182–197, 2002, doi: 10.1109/4235.996017.
10. S. Jha and V. K. Jain, "Design and development of the magnetorheological abrasive flow finishing (MRAFF) process," *International Journal of Machine Tools and Manufacture*, vol. 44, no. 10, pp. 1019–1029, Aug. 2004, doi: 10.1016/J.IJMACHTOOLS.2004.03.007.
11. A. Sidpara and V. K. Jain, "Rheological properties and their correlation with surface finish quality in MR fluid-based finishing process," *Machining Science and Technology*, vol. 18, no. 3, pp. 367–385, Jul. 2014, doi: 10.1080/10910344.2014.925372.
12. A. S. Rajput, M. Das, and S. Kapil, "A comprehensive review of magnetorheological fluid assisted finishing processes," *Machining Science and Technology*, vol. 26, no. 3. Taylor and Francis Ltd., pp. 339–376, 2022. doi: 10.1080/10910344.2022.2129982.
13. S. Karthikeyan, B. Mohan, S. Kathiresan, and G. Anbuchezhiyan, "A study on in-vitro behaviour of stainless steel 316L subjected to rotational magnetorheological abrasive flow finishing process," *Surf Topography*, vol. 10, no. 3, Sep. 2022, doi: 10.1088/2051–672X/ac9150.
14. A. Singh Rajput, S. Kapil, and M. Das, "Computer-aided process planning system for super finishing of flat surfaces with pockets through magnetorheological finishing process," *International Journal of Computer Integrated Manufacturing*, pp. 1–17, Mar. 2023, doi: 10.1080/0951192x.2023.2189313.
15. V. Kumar, R. Kumar, and H. Kumar, "Development and investigation of the magnetorheological abrasive finishing (MRAF) system to nano-finish exterior cylindrical surfaces," *Materials and Manufacturing Processes*, 2022, doi: 10.1080/10426914.2022.2072881.
16. S. Karthikeyan, B. Mohan, S. Kathiresan, and G. Anbuchezhiyan, "Effect of process parameters on machinability, hemocompatibility and surface integrity of SS 316L using R-MRAFF," *Journal of Materials Research and Technology*, vol. 15, pp. 2658–2672, Nov. 2021, doi: 10.1016/J.JMRT.2021.09.060.
17. Y. Choopani, M. Khajehzadeh, and M. R. Razfar, "Experimental investigations into the nano-finishing of Al2024 tubes using the rotational-magnetorheological abrasive flow finishing (R-MRAFF) process," *Proceedings of the Institution of Mechanical Engineers, Part E: Journal of Process Mechanical Engineering*, vol. 236, no. 6, pp. 2545–2557, Dec. 2022, doi: 10.1177/09544089221094999.
18. M. Kumar, A. Kumar, H. N. S. Yadav, and M. Das, "Gear profile polishing using rotational magnetorheological abrasive flow finishing process," *Lecture Notes in Mechanical Engineering*, pp. 565–576, 2023, doi: 10.1007/978–981-19–3266-3_44/COVER.

19. V. Kumar and S. Chakraborty, "Analysis of the surface roughness characteristics of EDMed components using GRA method," *Proceedings of the International Conference on Industrial and Manufacturing Systems (CIMS-2020)*, 2022, pp. 461–478. doi: 10.1007/978-3-030-73495-4_32.
20. A. Sharma, V. Kumar, A. Babbar, V. Dhawan, K. Kotecha, and C. Prakash, "Experimental investigation and optimization of electric discharge machining process parameters using grey-fuzzy-based hybrid techniques," *Materials*, vol. 14, no. 19, p. 5820, Oct. 2021, doi: 10.3390/ma14195820.
21. V. Kumar, P. P. Das, and S. Chakraborty, "Grey-fuzzy method-based parametric analysis of abrasive water jet machining on GFRP composites," *Sādhanā*, vol. 45, no. 1, p. 106, Dec. 2020, doi: 10.1007/s12046-020-01355-9.
22. V. Kumar, S. Diyaley, and S. Chakraborty, "Teaching-learning-based parametric optimization of an electrical discharge machining process," *Facta Universitatis, Series: Mechanical Engineering*, vol. 18, no. 2, pp. 281–300, 2020, doi: 10.22190/FUME200218028K.
23. S. Chakraborty, P. P. Das, and V. Kumar, "Application of grey-fuzzy logic technique for parametric optimization of non-traditional machining processes," *Grey Systems: Theory and Application*, vol. 8, no. 1, pp. 46–68, 2018, doi: 10.1108/gs-08–2017–0028.
24. S. Bandaru, R. Tulshyan, and K. Deb, "Modified SBX and adaptive mutation for real world single objective optimization," *2011 IEEE Congress of Evolutionary Computation, CEC 2011*, pp. 1335–1342, 2011, doi: 10.1109/CEC.2011.5949771.

3 Nanofinishing of metallic mirrors

Munish Kumar, Ajay Choudhary, and Dilshad Ahmad Khan

3.1 INTRODUCTION

Finishing is defined as a process which alters the surface of a workpiece to achieve the desired appearance or to remove machining marks and provide durability. Manufacturers often use a variety of finishing processes to improve their products. Without finishing, the lifespan of the products would be reduced due to corrosion and wear, and their function would be diminished. Some industrial applications require a very high level of surface finish and precision. The surface qualities needed to meet the demand for high precision and finish cannot be produced by conventional methods of finishing (like grinding, lapping, etc.). Manufacturers all across the world realized that high-precision manufacturing was necessary to increase accuracy and promote micro- and nanotechnology.

The categorization of machining processes is based on the level of accuracy achieved. These processes can be divided into three categories: standard, precision, and ultra-precision processes. The term "ultra-precision machining" describes techniques that have attained or are capable of attaining the maximum level of dimensional accuracy at a specific time [1]. In conventional machining, the achieved accuracy typically reaches 1 μm, while 0.01 μm (10 nm) and 0.001 μm (1 nm) of accuracy can be obtained through precision and ultra-precision machining, respectively.

Common Conventional Finishing Processes:

a. **Grinding:** The most commonly used method of finishing is grinding. A grinding wheel is used to grind the metal surface in order to remove any flaws or scratches. This can be done using a variety of abrasive materials.
b. **Polishing:** Polishing is a surface finishing process and is used to get a reflective surface. The polishing process follows the grinding. The metal surface is polished using a series of finer and finer abrasive materials until the desired level of smoothness and shine is achieved. Polishing compounds can be used to speed up the process and achieve a more uniform finish.
c. **Buffing:** Once the surface has been polished, it is buffed to further enhance the shine and remove any remaining imperfections. This can be done using a buffing wheel or pad and a polishing compound.

DOI: 10.1201/9781003496298-3

d. **Electrolytic polishing:** Electrolytic polishing is a process that uses an electric current to selectively remove material from the surface of the metal. This can be used to remove any remaining surface defects and create an even smoother, more reflective finish.
e. **Vapour polishing:** Vapour polishing is a process that involves exposing the metal to a vapour of a polishing agent, which then condenses on the metal surface, dissolving any surface defects. This process can be used to create a highly uniform, mirror-like finish on complex geometries.

3.2 NANOFINISHING

Nanofinishing processes refer to ultra-precise finishing techniques that have been created to achieve surface finishes at the nanometre level [2]. Initially, the term "nanofinishing" was used to classify machine systems and integrated manufacturing technologies capable of extremely precise machining with an accuracy of around 1 nm. These procedures have been developed over the past few decades to get around some of the drawbacks of traditional finishing methods, such as the need for harder tools and the ability to control finishing forces precisely during operation.

By addressing these limitations, nanofinishing processes have enabled the finishing of harder materials and improved control over the final surface characteristics during the manufacturing process. Additionally, advanced finishing processes (AFPs) utilizing loose abrasives have relaxed the constraints on finishing complex geometries by allowing abrasives to reach difficult-to-reach areas of a workpiece surface. Consequently, these recently developed finishing processes have significantly contributed to achieving the requirements of twenty-first-century manufacturing.

The demand for nano-finished materials and products has surged across various industries, including aerospace, automobile, biomedical, and defence, among others.

3.3 NANOFINISHING OF METALLIC MIRRORS

Metal mirrors have been used by humans for at least 6000 years. But after it was discovered that glass with a small layer of metal is highly reflecting and, in addition to other advantages, more scratch resistant, the use of metal mirrors started to decline. Copper mirrors, with their highly reflective surfaces, are believed to have been among the earliest metal reflectors used by ancient civilizations. Their discovery and utilization marked a significant advancement in human history, enabling the manipulation and redirection of light for practical and aesthetic purposes.

In the past century, metal mirrors made a remarkable comeback, driven by their distinct advantages that proved invaluable for a range of new and intriguing applications. One notable application involved using metal mirrors as reflectors in laser systems. Due to their high thermal conductivity, these mirrors efficiently dissipated heat generated during laser operations, ensuring optimal performance and preventing any potential damage. This thermal conductivity property became a crucial factor in enhancing the overall efficiency and stability of laser applications.

Additionally, metal mirrors found utility in the field of telescopes. Their ability to bend and manipulate light beams played a vital role in shaping the optical path within telescopic systems. By precisely controlling the direction and focus of light, metal mirrors contributed to improving the accuracy and clarity of astronomical observations. This advancement opened up a plethora of opportunities for astronomers and researchers to explore the mysteries of the universe with enhanced precision and depth.

The emergence of metal mirrors in these unique applications brought forth a wide range of new possibilities. Scientists, engineers, and innovators eagerly explored and capitalized on the specific properties offered by metal mirrors, ushering in an era of innovative technologies and scientific breakthroughs. Whether it was the efficient dissipation of heat in laser applications or the precise manipulation of light in telescopes, metal mirrors proved themselves as indispensable tools, revolutionizing their respective fields and inspiring further advancements.

Metal mirrors are widely utilized in optical applications, and three common materials for constructing them are aluminium, copper, and beryllium. Each material has its unique characteristics and benefits. Copper's high thermal conductivity makes it especially desirable for laser applications. This means that it can effectively dissipate heat generated by the laser, making it a suitable choice for such purposes. Aluminium, on the other hand, offers good reflectivity even at shorter wavelengths. It is also lightweight and easily malleable, similar to beryllium. These properties make aluminium a versatile material for constructing metal mirrors that can be used in various optical applications. Beryllium is known for its stiffness and light weight, which are advantageous qualities for constructing metal mirrors. However, it has a significant drawback: it exhibits toxic behaviour. Due to this toxic nature, beryllium is not commonly used, despite its desirable mechanical properties. Safety concerns surrounding beryllium restrict its widespread usage in optical applications.

To achieve the required precision, the manufacturing process of metallic mirrors involves several consecutive steps, each serving a specific purpose. Here is an overview of the steps involved:

i. ***Prefabrication:*** The initial step involves shaping the metal material into the desired form. This can be done through techniques like machining, milling, turning, or casting, depending on the complexity of the component.
ii. ***Stress relieving:*** Metal optics are prone to internal stresses induced during the fabrication process, which can lead to deformation or warping. Stress relieving techniques, such as annealing or heat treatment, are applied to reduce these internal stresses and ensure dimensional stability.
iii. ***Plating:*** Metal optics may undergo plating, which involves depositing a thin layer of another metal or alloy onto the surface. Plating can enhance the optical properties, protect against corrosion, or provide a suitable surface for subsequent processing steps.
iv. ***Figuring:*** Figuring is a crucial step that involves shaping the metal optic to achieve the desired optical surface form. It requires precision grinding, polishing, and sometimes diamond turning or lapping techniques. This process helps achieve the required surface smoothness and curvature for optimal light transmission or reflection.

v. ***Finishing:*** Once the desired shape and surface roughness are achieved, additional finishing steps may be performed to enhance the surface quality further. This can include processes like honing, buffing, or chemical treatments to remove any remaining imperfections, scratches, or contaminants.
vi. ***Optical coating:*** Metal optics are often coated with thin layers of specialized materials to enhance their optical performance. Optical coatings can reduce reflections, increase transmission, or provide specific spectral properties. Common coating techniques include vapour deposition or sputtering.

Throughout these consecutive processing steps, precision and attention to detail are crucial. The use of advanced machinery, measurement tools, and skilled operators is necessary to ensure the final product meets the required specifications for shape, roughness, and optical performance. The overall goal is to produce metal optics that provide accurate and high-quality light manipulation for various optical applications.

Numerous advancements are occurring in the field of material finishing using fine abrasives. These advancements encompass various aspects such as the processes involved, the types of abrasives utilized, and their bonding techniques. These developments enable the achievement of surface finishes at the nanometre scale [2].

Alam et al. [3] investigated the subject and discovered that in the past there existed a constraint on the size of abrasives, typically around a few micrometres. However, recent advancements in materials synthesis have now made it possible to manufacture abrasives in the nanometre scale, allowing for ultra-fine sizes. Abrasives are employed in different ways, including coated abrasives, bonded abrasives, such as grinding wheels, and loose abrasives like those used for lapping and polishing, all serving the purpose of creating components with diverse shapes, sizes, precision, smoothness, and surface integrity.

When the size of chips becomes extremely small, reaching atomic dimensions of around 0.3 nm, achieving the highest level of accuracy through finishing processes is possible. To achieve surface finishes at the nanometre scale, it becomes necessary to remove material by individually removing molecules or atoms, either separately or in clusters. Certain techniques, such ion beam machining (IBM) and elastic emission machining (EEM), perform by directly removing atomic particles and molecules from the material's surface. In contrast, alternate methods that rely on abrasive machining eliminate them in groups.

Depending on the type of energy they require, AFPs can be categorized as mechanical, thermoelectric, electrochemical, or chemical. The effectiveness and application of each particular process are influenced by the material of workpiece characteristics and the particular functional specifications of the component.

Precise control of finishing forces is crucial in mechanical AFPs. Numerous recently developed AFPs utilize magnetic or electric fields to control finishing forces on abrasive particles externally. The field-controlled AFPs demonstrate significant potential for nanofinishing in metal optics, benefiting industries such as lasers, astronomical instruments, and consumer electronics. These sectors frequently employ metallic mirrors and other optical components to enhance system reliability and operational availability [1]. Due to its exceptional qualities such as high reflectivity, good thermal conductivity, durability, and lightweight nature, the utilization of metallic mirrors becomes indispensable in the case of high-power multi-kilowatt lasers. Mirror finishing processes are versatile techniques employed on different metals like aluminium and copper, among others. The specific method chosen relies on factors like the metal type, the desired outcome, and the intended purpose of the final product. By employing appropriate processes, these metals can be transformed into reflective surfaces with a mirror-like finish, catering to various industrial or aesthetic requirements.

Due to its high reflectance and lightweight nature, aluminium is a popular option for metallic mirrors. It surpasses all other metals in terms of reflectance in the ultraviolet and infrared regions, second only to silver in the visible light and near-infrared ranges [4]. Copper, which ranks as the third most utilized metal globally, finds extensive applications in various domains such as integrated circuits (IC), aircraft, ultra-precision machining, electric power, catalysis, and more. This is attributed to its beneficial features like low resistivity, excellent electrical conductivity, and so on. When employing copper devices, it is imperative to ensure exceptional surface quality [5]. Titanium is an important metal due to its light weight. It has good corrosion resistance, mechanical characteristics, and biocompatibility [6]. Pure nickel has excellent physical and mechanical properties such as high corrosion resistance, high thermal strength, low friction coefficient, high hardness, strong wear resistance, non-magnetic behaviour, good biocompatibility, and high electro-catalytic activity [7].

3.4 ADVANCED MIRROR FINISHING PROCESSES

Advanced mirror finishing processes refer to a set of techniques employed to attain a superior level of shine and reflectivity compared to conventional mirror finishing methods. These advanced processes are particularly utilized in specialized industries that demand extremely high precision and reflectivity, such as aerospace, optics, and other advanced technological fields.

1. ***Single point diamond turning (SPDT):*** This technique involves removing material from the workpiece surface using a diamond tool, producing a finish that is extremely precise and smooth. The SPDT tool is controlled by a computerized system (CNC machine), allowing for precise control of the cutting parameters and achieving exceptional surface quality.
2. ***Chemical mechanical polishing (CMP):*** In CMP, the workpiece is polished using both chemical and mechanical processes in a combination. A slurry consisting of abrasives and chemical agents is applied to the workpiece, and a polishing pad or wheel is used to generate relative motion between the pad and workpiece. This process helps remove surface imperfections and obtain a highly reflective finish.
3. ***Electrochemical finishing:*** This process utilizes the principle of electrochemistry to achieve mirror-like surfaces. It involves immersing the workpiece in a solution of electrolyte and applying an electric charge. The controlled electrochemical reactions help in selectively removing material and smoothing out the surface, resulting in a high-gloss finish.

4. ***Magnetic assisted finishing:*** Magnetic fields are employed to precisely polish the surface while controlling the finishing forces for material removal. The workpiece is immersed in a magnetic fluid and subjected to a magnetic field that generates controlled forces on the surface. This method allows for accurate control over the finishing process and enhances surface smoothness.
5. ***Laser finishing:*** Laser beams are used to modify the properties of workpiece surface in this process. The laser energy is carefully controlled to melt or vapourize the material, which helps in removing surface irregularities and achieving a polished surface. Laser finishing offers excellent precision and can be used for complex geometries.

These advanced mirror finishing processes enable industries to achieve exceptional levels of shine, smoothness, and reflectivity in their products. The use of these techniques is crucial in applications where optical clarity, precision, or reflective surfaces are essential for optimal performance.

The processes mentioned earlier have been thoroughly discussed in the subsequent subsections:

3.4.1 Single point diamond turning

The single point diamond turning abbreviated as SPDT is a method of precise machining used to remove materials on a very small scale, specifically in the nanometre range. It involves using diamond tools to cut and shape the material being machined. The capability of diamond turning to produce machined surfaces with excellent optical quality is one of its main benefits. The surfaces created through diamond turning have a high level of smoothness and accuracy, which is crucial in applications where precise optics are required [8]. For the creation of aspherical and diffractive surfaces, diamond turning is especially well suited. Diffractive surfaces are designed to manipulate light waves, and aspherical surfaces have a non-spherical shape. Both types of surfaces are commonly used in optics and photonics applications.

The diamond tool used in the SPDT process possesses two attractive features: nanometric edge sharpness and wear resistance. The nanometric edge sharpness refers to the extremely small cutting edge of the diamond tool, which allows for the precise material removal at the nanometre scale. Wear resistance means that the diamond tool can maintain its sharpness and cutting performance for an extended period without significant deterioration.

By utilizing diamond tools in the SPDT process, it becomes possible to achieve sub-nanometre surface finishes on metal mirrors. The resulting mirror surfaces are exceptionally smooth at the atomic or molecular level, allowing for high-quality reflection of light.

Diamond tools undergo a grinding and lapping process to achieve a cutting-edge radius in the sub-micron range. However, tool cutting edge cannot be regarded as completely sharp due to the lower relationship between the tool nose radius and undeformed chip thickness. Ultra-precision machining demonstrates superior capabilities compared to conventional processes. The quality of the machined surface in SPDT is primarily affected by parameters such as the temperature distribution between the tool tip and the workpiece, the cutting parameters, the tool geometry, and the orientation of the crystals on the workpiece. Like traditional machining, the diamond turning process involves the removal of material by moving the tool and workpiece together to achieve the desired shape and amount.

Diamond tools are carefully sharpened through a process of grinding and polishing to create an extremely fine cutting edge. However, despite this sharpness, the tool cutting edge is not perfectly sharp because the relationship between the radius of the tool's edge and the material thickness being cut is not proportional. Ultra-precision machining, which includes processes like diamond turning, is highly capable compared to traditional methods. In diamond turning, the cutting parameters (like speed, feed, and depth of cut) and tool shape have the greatest impact on the quality of the surface being machined, the orientation of the crystal in the material being worked on, and the distribution of heat between the tool and the workpiece. Like conventional machining, diamond turning involves removing material from the workpiece to achieve a desired shape by moving the workpiece and the tool together.

In a SPDT machine, a fixture secures the workpiece and it is pulled towards a vacuum chuck using vacuum pressure. The precise rotation of the spindle in this ultra-precision machine is accomplished with air bearings. The movement of the machine's axis is measured by a laser interferometer or a linear encoder, and the diamond tool is controlled by numerical commands. To maintain ideal conditions, diamond turning machines are housed in a controlled environment free of dust and with regulated temperature. Figure 3.1 depicts the SPDT machining system, which is composed of a vacuum chuck, a diamond tool, and a fixture attached to an aerostatic spindle. The diamond turning process is a dependable method for creating free-form optics, metal mirrors for astronomical telescopes, and reflective mirrors.

In the study conducted by Li et al. [9 , they investigated SPDT and its ability to produce highly reflective surfaces to create extremely precise mirrors from Al6061, a type of aluminium alloy. The goal was to achieve a mirror finish with a roughness lower than 10 nm and with errors lower than 200 nm. The researchers performed

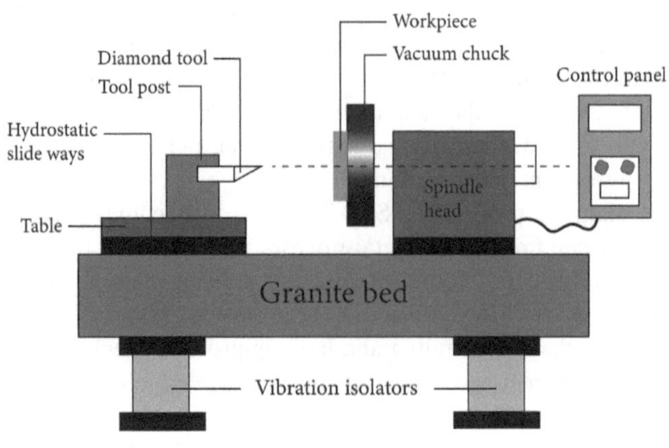

FIGURE 3.1 Machining system of SPDT.

their experiments on an Al6061-T6511 workpiece using an ultra-precision machine 450UPL (Moore Nanotech). Fifteen samples were tested, and, during each test, spindle rotational speed and depth of cut remained fixed at 1000 RPM and 5 micrometres respectively. However, the feed rate, which represents the speed at which the tool moves along the surface, varied from 1 to 15 mm per minute for each sample. To measure the surface roughness, a white light interferometer (Zygo Nexview) was employed, which could analyse an area of 700×700 micrometres. The researchers found that the measured reflectance values were inconsistent at feed rates lower than 7 mm/min, varying between 84% and 90%. Surface roughness dropped from 7 to 5 nm as the feed rate drop even more, from 3 to 1 mm/min. On the other hand, when the feed rate exceeded 7 mm/min, the surface roughness decreased. This smoother surface led to an increase in reflectance. However, once the rate of feed was reduced from 7 to 3 mm/min, the surface roughness continued to reduce, but the reflectance stabilized at the same level. Chip formation during the machining process became unstable when the feed rate dropped lower than 3 mm/min, resulting in a decrease in reflectance, which was unfavourable for achieving the desired mirror-like finish.

Zhang et al. [10] investigated the material removal mechanism and surface creation of SPDT. Four quadruple-symmetric fan-shaped patterns were formed by the Cu (100) crystal plane, four double-symmetric fan-shaped patterns by the Cu (110) crystal plane, and six triple-symmetric fan-shaped patterns by the Cu (111) crystal plane. Molecular dynamics and cutting dynamics were integrated into a dual dynamic model to present the nano-surface development in single-crystal copper SPDT. Cutting dynamics modelling and molecular dynamics modelling are joined by the friction coefficient, which varies with atom configuration. The Cu (100) crystal plane is the best candidate for machining out of the three single-crystal (110), (111), and (110) crystal planes because it produces the smoothest surface at a value of 15.433 nm. The next crystal plane is Cu (111), with a surface roughness measurement of 17.016 nm. The greatest surface roughness, measuring 20.102 nm, was found on the Cu (110) crystal plane.

The study conducted by Zong et al. [11] is on the importance of diamond turning in the fabrication of precision components in a variety of industries, including optics and renewable energy. In their experiments, the researchers used a machine tool called Nanoform 250 (manufactured by Taylor Hobson PNEUMO) and employed a diamond tool. Pure copper plates were selected as the material for the workpieces. The process parameters were as follows: 3000 rpm spindle speed, 3 mm nominal cutting depth, and variable feeding velocity with varied feed rates. The experiments were conducted using four different tools with varying tool radii: Tool 1 had a radius of 1.856, Tool 2 had a radius of 0.954, Tool 3 had a radius of 0.495, and Tool 4 had a radius of 0.045. The researchers observed that Tool 3 or Tool 4 provided a higher surface roughness than Tools 1 and 2. This finding supports the idea that the elastic recovery and the minimum undeformed thickness of chip are dependent on the cutting-edge radius in diamond turning. The surface roughness values achieved using the different tools were as follows: Tool 1 produced a roughness value of 29 nm, Tool 2 a roughness value of 25.5 nm, Tool 3 a roughness value of 46.5 nm, and Tool 4 a roughness value of 15 nm. Therefore, by adjusting the process parameters and tool geometries based on a quantitative ratio of the maximum undeformed thickness of chip to the radius of cutting edge, it is possible to maintain the desired surface finish.

3.4.2 Chemical mechanical polishing

In the interest of increasing the material removal rate (MRR), chemical as well as mechanical processes are employed in CMP. By employing CMP, it becomes possible to achieve a flat, smooth, and undamaged surface [12]. In the industry, there are four main types of CMP equipment that are widely utilized and considered representative. These include: (a) a rotary-type polisher equipped with a wafer carrier that moves back and forth along the diameter of the platen, (b) a rotary-type polisher with an oscillating carrier, (c) an orbital-type polisher with an orbitally rotating platen, and (d) a linear-type polisher that uses a linear motion belt for polishing purposes.

In the semiconductor industry, a standard tool called a rotary CMP tool is utilized to achieve surface finishing on wafers. This tool comprises a rotating plate known as a platen and a wafer carrier. Both the wafer and platen carrier rotate in synchronization, with the wafer carrier also moving back and forth along the platen's radial

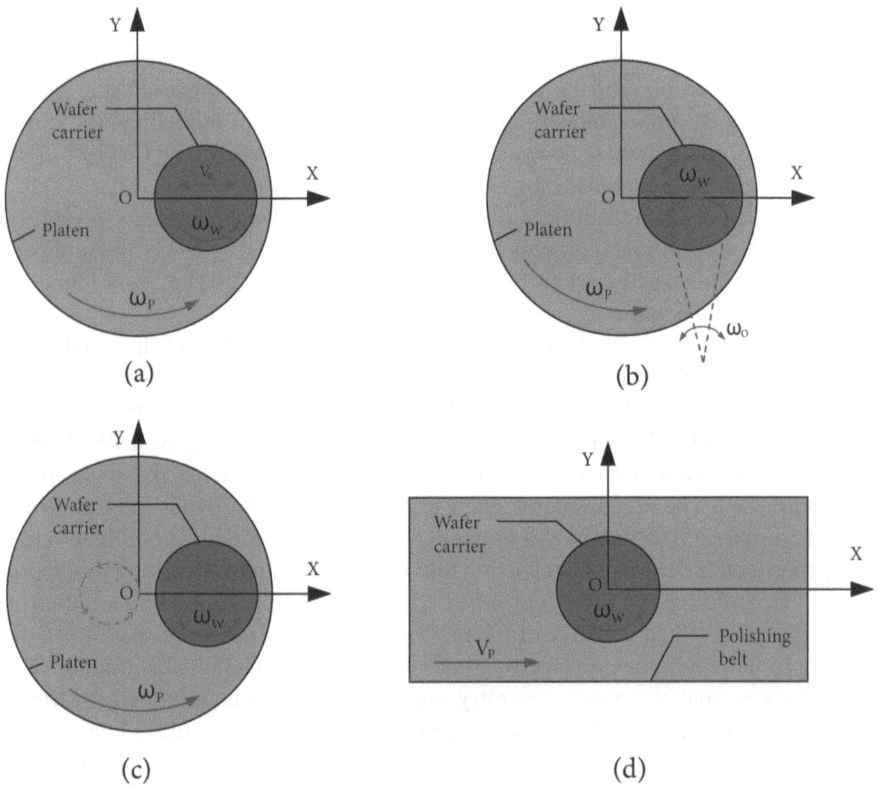

FIGURE 3.2 Schematic of CMP equipment's type: (a) rotary type with a reciprocation mode; (b) rotary type with an oscillation mode; (c) orbital type; and (d) linear type.

direction. The wafer, which is the item being polished, is held inside the revolving carrier, and a polishing pad is secured to the rotating disc. The required relative motion for the polishing process is generated by the movement of the platen and the carrier together. To facilitate the finishing procedure, a slurry consisting of chemical solutions and particles is applied to the polishing pad, as depicted in Figure 3.3. These particles, ranging in size from 10 to 100 nanometres on average, serve as abrasive agents.

As a consequence of the interplay between the mechanical activities of the abrasives and the chemical actions of the solutions, very little amounts of material are removed from the wafer's surface while polishing. This allows for precise surface finishing to be achieved, which is crucial in the manufacturing of semiconductor devices.

To enhance the consistency of CMP for large-size wafers, a new multizone carrier is being developed. Properly designed polishing pads with grooves contribute to effective polishing. CMP relies on a complex consumable called slurry, which is a stable combination of abrasive, deionized water, and other chemicals. The balance and interaction of oxidants, inhibitors, and complexing agents significantly impact the polishing outcome. While CMP can achieve nanoscale finishes, it is only capable of processing flat surfaces. An ideal nano finishing process involves removing a small amount of material. In an abrasive slurry-based process, each abrasive particle should ideally create atomic or nanoscale surfaces. Additionally, it is crucial to prevent vibration transfer from the tool to the work surface. CMP, or EEM, possesses all these desirable features. It can also machine surfaces with various profiles by controlling the time of machining and tool motion.

Zhao et al. [12] investigated the application of CMP for achieving several desirable outcomes in surface finishing. Zhao et al. found that by employing CMP, they were able to obtain a planar surface, meaning a surface that is flat and level. This is important in many applications where surface uniformity is required, such as integrated circuit fabrication or optical component manufacturing. The planarization achieved through CMP ensures that the entire surface is at the same level, eliminating any height variations or steps that may be present.

Additionally, the authors stated that CMP was capable of producing a "nanofinishing" effect. Nanofinishing refers to the achievement of extremely fine surface finishes at the nanoscale level. This is crucial in various high-precision applications,

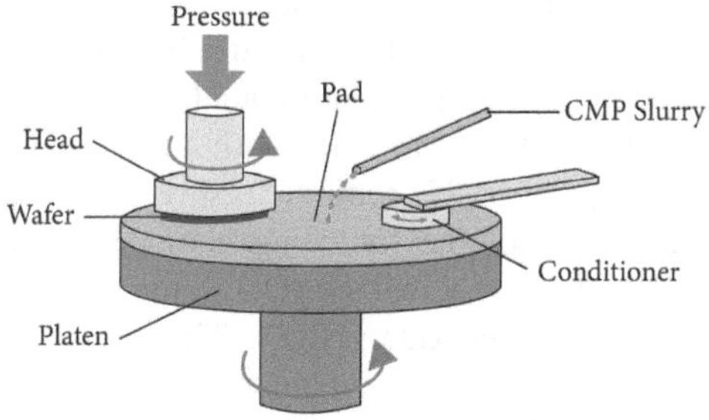

FIGURE 3.3 Schematic diagram of CMP process.

including nanoelectronics, optoelectronics, and advanced sensors. The ability of CMP to create such nanoscale finishing allows for enhanced performance, improved optical properties, and increased functionality of the finished surfaces. Overall, the research conducted by Zhao et al. [12] demonstrated that CMP is an effective method for obtaining planar, smooth, and damage-free surfaces. The technique offers advantages in terms of surface uniformity and the ability to achieve nanofinishing, making it a valuable process in industries that require high-precision surface characteristics.

Joshi et al. [13] explored that CMP is capable of achieving nanofinished surfaces. The study utilized a setup comprising stacked *X-Y* stages and a decoupled *Z* stage. The workpiece was placed within an acrylic tank filled with abrasive slurry, which was mounted on a dynamometer attached to the *X*-coordinated three-axis motion. The investigation considered three types of independent variables: (1) process condition variables, (2) abrasive slurry variables, and (3) tool stiffness variables. The major polishing parameters were identified as the rotational speed and applied load, which influenced finishing of surface by affecting polishing pressure and thickness of hydrodynamic film. The stiffness of the tool was also examined as it could impact polishing pressure in the hydrodynamic film. Polishing was conducted for different durations: 3, 13, 23, and 35 minutes. With increased polishing time, the surface roughness increased in response to higher load. However, the average surface roughness decreased when the size of the abrasive increased from 0.3 to 1 μm over time (specifically at 23 and 35 minutes). Optimal spindle speed for achieving the smoothest surface was found to be 2400 rpm at shorter polishing durations. CMP demonstrated the ability to produce nano-finished surfaces on hard materials, with the best average surface roughness (Ra) value achieved at 3.5 nm. A slurry containing 0.3 μm abrasive particles yielded the best surface finish for shorter finishing times (up to 13 minutes), while a slurry with 1 μm abrasive particles resulted in the best surface finish for longer finishing times (after 35 minutes). Additionally, the surface finish improved when all polishing parameters were set to their minimum levels, including minimum abrasive concentration (0.05 μm size of particle), minimum stiffness of tool, minimum load (1 N), and minimum rotational speed (1000 rpm).

Kaushik et al. [6] investigated the use of titanium (Ti) and its alloys as important materials for achieving mirror-finish surfaces through CMP. These materials are valued for their favourable mechanical properties, lightweight nature, and biocompatibility. In the study, a Ti alloy workpiece with dimensions of 15 mm^3 was employed. The workpieces were securely held in Bakelite moulds during the polishing process. Mechanical polishing was carried out using a sequence of silicon carbide (SiC) abrasive papers, followed by cleaning with ethanol and subsequent drying prior to CMP.

The polishing procedure was carried out manually using a semi-automatic polishing machine (LBQ060, Buehler) furnished with a high-quality polishing fabric, operating at a rotational speed of 233 rpm. A freshly prepared polishing solution was utilized, which consisted of the following components: (i) 4 vol% emulsifiable water-soluble oil, (ii) 80 vol% deionized (DI) water, and (iii) 16 vol% of a 30% aqueous solution of hydrogen peroxide (H_2O_2). The resulting solution was subjected to ultrasonication for 10 minutes at room temperature. Observations revealed that after 10 minutes of polishing, the workpiece surface exhibited a mirror-like surface, shiny appearance with no visible scratches under optical micrographs.

Analysis of the optical profiles of the polished surface indicated a surface flatness of approximately 0.90 μm, with a 2.33 μm of maximum height. Furthermore, the root-mean-square (RMS) roughness (Rq) and arithmetic mean deviation of roughness (Ra) were determined to be 1.12 nm and 0.96 nm, respectively.

Zhang et al. [7] explored the properties of a novel environmentally friendly CMP slurry that comprised silica, malic acid, deionized water, H_2O_2, and nickel. The slurry exhibited excellent mechanical, physical, and chemical characteristics. The authors measured the peak-to-valley (PV) and average surface roughness (Ra) values to be 4.49 nm and 0.44 nm, respectively, using a scan area of $71 \times 53\,\mu m^2$. Their findings revealed that pure nickel and its alloys exhibited the lowest surface roughness among the tested materials at the specified scan area. To understand the CMP mechanism, the researchers employed infrared (IR) measurements. Initially, H_2O_2 facilitated the oxidation of the nickel alloy, leading to the formation of nickel, chromium, and molybdenum oxides on the surface. Following that, malic acid subsequently dissolved the nickel oxides while leaving the chromium and molybdenum oxides unaffected. The introduction of Ni_2+ and Ni_3+ ions into the CMP slurry, with malic acid chelated with these ions, prevented the discharge of nickel ions into the environment. Finally, the mechanically induced action of CMP removed the generated oxides.

3.4.3 Electrochemical finishing

Electrochemical polishing is a technique that is used to create a smooth and reflective surface on metals and alloys. It is achieved through an anodic treatment, which means applying a positive electric charge to the metal. This method is highly effective in achieving a mirror-like finish on metal surfaces [14]. The success of electrochemical polishing depends on various elements, including the properties of the metal's characteristic, the composition of the electrolyte (the liquid in which the process takes place), temperature, voltage, current density, and time. All of these variables interact to determine the process's final conclusion. During anodization, different changes occur on the metal surface, such as etching (controlled removal of material), pitting (formation of small holes), deflashing (removal of excess material), the formation of a passive film (protective layer), self-organization of nanostructures (arrangement of very small particles), deburring (removal of rough edges), trimming (reshaping), and smoothing [14].

When using anodic polarization for electropolishing, the choice of the electrolyte becomes crucial. It must have specific chemical properties to achieve the desired results. When trying to create mirror-like finishing on metals or alloys, the use of aqueous solutions is limited. This is because water can react with the electric current, leading to the generation of molecular oxygen instead of the desired polishing effect, Eq. (3.1)

$$2H_2O - 4e^- \rightarrow O_2 + 4H^+ \tag{3.1}$$

Although water is not as reactive as base metals, its reactivity in aqueous solutions is still significant. In the context of electrochemical polishing, the evolution of oxygen during the process leads to the formation of secondary metal oxide films. Only a few metals can be polished to have a glossy surface in aqueous acid solutions. These metals need to be resistant to acids and have thin passive oxide films on their surface, such as stainless steel, nickel, copper, platinum, and similar materials. However, for most significant metals for industry, anodization in acid solutions results in pitting or etching rather than achieving a mirror-like surface. To understand the fundamental process, it is necessary to consider the electrochemical oxidation of water that takes place on the surface of metal. The pores of the oxide layer were penetrated by water molecules (H_2O). H^+ ions were consequently electrochemically generated and moved in the opposite direction, towards the cathode, as a result. These hydrogen ions can react with the metal oxide and the rate at which they migrate to the cathode as well as the oxide layer thickness determine the likelihood of the heterogeneous chemical reaction, which leads to the dissolution of the oxide layer, Eq. (3.2). This dissolution of the oxide film is a critical factor for achieving a mirror-like metal surface as it helps to reduce the diffuse scattering of light.

$$M_xO_y + 2yH^+ = xM^+(2y/x) + yH_2O \tag{3.2}$$

Equation (3.2) represents a chemical reaction involving a metal oxide (M_xO_y), hydrogen ions (H^+), metal ions (M^+), and water (H_2O). The reaction shows the transformation of the metal oxide and hydrogen ions into metal ions and water [14].

The probability of this reaction occurring, as described in Eq. (3.2), decreases when the rate of migration of hydrogen ions (H^+) to the cathode (positive electrode) increases. The rate of H^+ migration is affected by a number of elements, including the viscosity of the electrolyte (the medium in which the reaction occurs), the porosity, thickness, and chemical characteristics of the oxide layer, as well as current density, temperature, voltage, and other variables. In the context of electrochemical processes, ionic liquids have gained significant attention as electrolytes. Ionic liquids are a type of non-aqueous electrolyte that consists of ions. They possess unique properties that make them advantageous compared to aqueous solutions. These advantages contribute to the growing interest and increased utilization of ionic liquids in various electrochemical processes.

The electrolyte used in electropolishing metals and alloys greatly influences the process due to its chemical nature. Various industries, such as microelectronics, space technology, aerospace construction, medical instruments, and biocompatible alloys, require highly reflective metal surfaces resembling mirrors. Electropolishing serves a similar purpose to the dissolution process, where an electric current is passed to remove the roughness of a surface and obtain a smooth surface. Outer layers of base metals and alloys consist of natural oxide films with varying compositions, covering imperfections in the crystalline structure. The attainment of a brilliant reflective surface involves a series of intricate steps, including the removal of metal oxides and controlled elimination of surface defects.

The efficiency of electropolishing relies on the chemical characteristics of the layer, which differ from one metal or alloy to the next. Generally, the mechanism of metal surface electropolishing involves the uneven removal rates of the protective films and the actual metal atoms from different areas of the surface. Equation (3.2) can be used to get rid of these oxide films in both aqueous and non-aqueous solutions.

In non-aqueous solutions, the oxidation of residual water leads to the generation of hydrogen ions. After the removal of the oxide coating, the edges of the crystal structure undergo surface smoothing at an increased current density. The removal of oxide films through chemical dissolution, as described in Eq. (3.2), is not the sole process responsible for their elimination. In cases where hydrogen ions are absent or scarce (such as in ionic liquids), the likelihood of electrochemically dissolving oxides increases, and they have the potential to undergo processes involving one or two electrons.

$$M_2O - 1e^- \rightarrow MO + M^+$$
$$MO - 2e^- \rightarrow M^{2+} + O_{ads}$$

especially for the low valence metals such as tin, copper, lead, etc.

Sun and Zou [15] devised a hybrid finishing approach that combines the advantages of magnetic abrasive finishing (MAF) and electrochemical machining (ECM). This study focused on investigating the finishing mechanism and finishing characteristics of the electrochemical magnetic abrasive finishing (EMAF) process. In their study, the EMAF technique was split into two finishing processes. The EMAF step serves as the first finishing step, while one MAF step serves as the second finishing step. The passive films formed via electrochemical reactions can simply polish the metal surface. The passive films can also be removed simultaneously by friction between the workpiece surface and a magnetic brush. The goal of the EMAF process is to create and remove passive films from the surface of the workpiece. It is undeniable that the EMAF process increases finishing efficiency over typical MAF by more than 75%, and the surface roughness is significantly reduced. The greatest experimental outcome of the EMAF procedure saw the surface roughness reduced from the initial value of 178 nm to less than 30 nm after 4 minutes in the EMAF stage, and it was finally reduced to 20 nm after 10 minutes in the MAF stage. In comparison to the MAF stage, about seven times as much material was eliminated during the hybrid finishing stage. Additionally, the EMAF process's effective finished area was only around 70% of the MAF process's.

By translating an electrolyte jet, Kawanaka et al. [16] have created unique electrochemical mirror finishing techniques, and the effects of translating speed and current conditions on surface roughness have been investigated. They looked at the three techniques for translating the electrolyte jet to mirror-finish the surface of arbitrary shapes. They used electrolyte jet machining process to mirror-finish the **stainless steel (SUS304).** The first technique makes use of a very fast jet translation speed. An expensive *XY* table with high feed speed and accuracy is required for this method. The second method has a low machining rate because it uses monopolar pulse current with a long interval and short duration. Bipolar pulses are proposed as a third method for speeding up machining. The surface roughness can be controlled by the current density (25 to 100 A/cm^2) in the electrolyte jet.

Regardless of the current density, it is feasible to reduce surface roughness when utilizing direct current by accelerating translation. The lowest surface roughness obtained at a translating speed of 100 mm/s was identical to that achieved without translation. Even when the translating speed was less than 1 mm/s, the use of monopolar pulse current with a short duration of 100 μs and a duty factor of less than 10% was able to achieve a mirror-like surface finish with Rz 0.2 μm. However, the rate of material removal was very slow. This led to the bipolar pulse current approach being proposed, and successful mirror finishing with duty factors above 50% was the result.

Maehata et al. [17] found that large workpieces made of stainless steel and regular carbon steel can be mirror-finished quickly and effectively using the electrolytic-abrasive method, producing surfaces with a roughness of up to 0.01 to 0.02 μm Rz. To produce the machining effect, they combined electrolysis with abrasive grains. With the help of grain action, metal is removed from this process mostly via electrolytic dissolution, which is proportional to working current and time, or subject to Faraday's law.

The anode metal is dissolved in an ionic unit when direct current and voltage are applied to two facing metals in an electrolytic solution. According to Faraday's law, the amount of the dissolved metal W is proportional to the amount of electricity applied $I \times t$ (current × time), expressed as

$$W = kIt\ (g)$$

where k is the electrochemical equivalent built into an anode material and η is the current efficiency used for dissolution and influenced by oxygen generation or oxide film formation; η is a proportion between a theoretical value and the amount of metal actually removed.

Wang et al. [18] explored a new stress-free machining process to create a copper (Cu) surface that is ultra-flat and ultra-smooth by combining electrochemical polishing (ECP) and jet electrochemical machining (Jet-ECM). With precise MRR manipulation and nozzle trajectory planning, the Jet-ECM process can effectively produce an ultra-flat surface. ECP using the same electrolyte as the Jet-ECM procedure can further reduce the workpiece's surface roughness. In accordance with the findings, the surface PV value, which represents the surface form error of the Cu surface, was decreased from 4.4 to 1.7 μm, and the surface roughness Ra was decreased from 70.3 to 13.5 nm. Jet-ECM and ECP, which use the same electrolyte and apparatus, can significantly reduce surface roughness and flatness.

3.4.4 Magnetic Assisted Finishing

In magnetic assisted finishing (MAF) processes, the goal is to improve the surface quality of a workpiece by applying precise finishing forces. Unlike traditional finishing methods where the forces are applied directly by mechanical means, MAF relies on the control of a magnetic field to exert these forces.

By controlling the distribution and strength of the magnetic field, the finishing forces applied on the workpiece can be precisely governed. This control allows for adjustments in the intensity of the finishing process, enabling fine-tuning of the surface quality achieved.

The advantage of using magnetic fields in finishing processes is that the forces can be controlled externally and adjusted as needed, providing a more flexible and controllable approach compared to traditional methods. This control allows for better precision, reduced damage to the workpiece, and the ability to achieve complex surface geometries and tolerances.

Overall, in magnetic assisted finishing (MAF) processes, the finishing forces are externally controlled by manipulating the magnetic field, leading to improved surface quality of the workpiece through selective material removal or polishing.

Following are the magnetic assisted finishing processes utilized for metallic mirror finishing:

3.4.4.1 Magnetic abrasive finishing (MAF)

In MAF, a workpiece is placed between two poles of a magnet, with one pole being North (N) and the other being South (S). Magnetic abrasive particles (MAPs) occupy the space between the workpiece and the magnet. This creates a magnetic field that forms a flexible brush called magnetic abrasive flexible brush (MAFB). The MAFB functions as a tool with multiple cutting points. The MAF process finds application in various industries, including automotive, aerospace, and medical device manufacturing, where precise finishing is necessary [19]. MAF allows for the finishing of complex shapes and hard-to-reach areas. It provides a range of surface finishes, from rough to highly polished. It is suitable for finishing different types of metals, such as steel, stainless steel, aluminium, and titanium. MAF is particularly advantageous when working on intricate geometries or parts that have challenging areas to access. Compared to other finishing methods, MAF is relatively fast and efficient. It can be easily automated for large-scale production. Furthermore, MAF generates less waste and has a lower environmental impact compared to traditional finishing techniques like grinding or polishing.

The process involves three main components: the workpiece, the magnetic fluid, and the magnetic field. The workpiece is typically made of a material that is difficult to finish using traditional methods, such as ceramics, glass, or certain metals. The

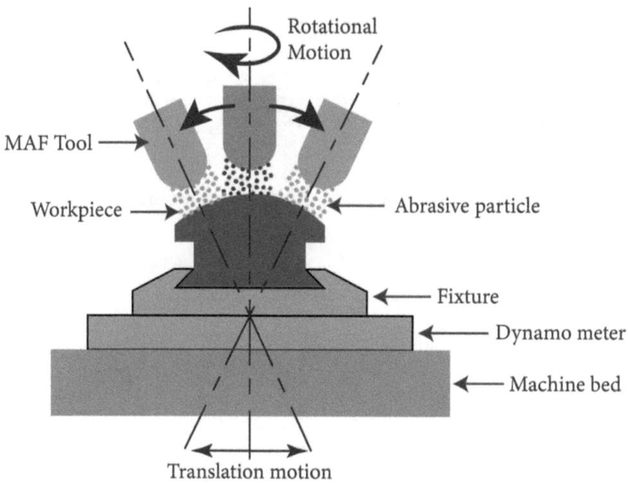

FIGURE 3.4 Magnetic abrasive finishing process.

magnetic fluid is a solution of MAPs in a liquid, such as water or oil. The magnetic field is generated using a magnetic coil or electromagnet.

In one of the researches on MAF on aluminium sheet, experiments were performed on a TN 50 vertical milling machine [4]. The North (N) and South (S) magnetic poles are positioned next to each other (Figure 3.5). In this arrangement, the grains of MAPs are attracted towards the magnetic poles, causing them to move away from the work surface rather than being compressed into it by magnetic forces, as shown in Figure 3.6.

As the magnetic poles rotate, the grains are pushed outward by centrifugal forces (as shown in Figure 3.7). When paired with the work surface's linear feed action, these grains effectively remove the high points or peaks on the surface. It can be visualized that larger grains with stronger magnetic forces are more effective at removing surface peaks, so increasing the size of MAP grains is advantageous for finishing flat surfaces. However, this is not the case for finishing cylindrical surfaces.

In this specific work carried out by Givi et al. [4], the chosen material is an aluminium alloy, which is widely used in the aerospace industry. The aluminium sheet being worked on has a thickness of 1 mm. The machine table reciprocates back and forth at a feed rate of 20.00 mm/min to provide a consistent finish throughout the whole workpiece surface. They carried out studies to determine the effects of various parameters on the finishing process and roughness of a surface. They observed that increasing the weight of flexible magnetic abrasive brush (FMAB) led to an increase in the number of particles engaged in the finishing process. This, in turn, resulted in an improvement in the roughness of the surface within a specific time of polishing. The rotational speed of the poles was found to have a significant influence on the roughness of surface improvement. Higher rotation speeds were associated with better surface quality.

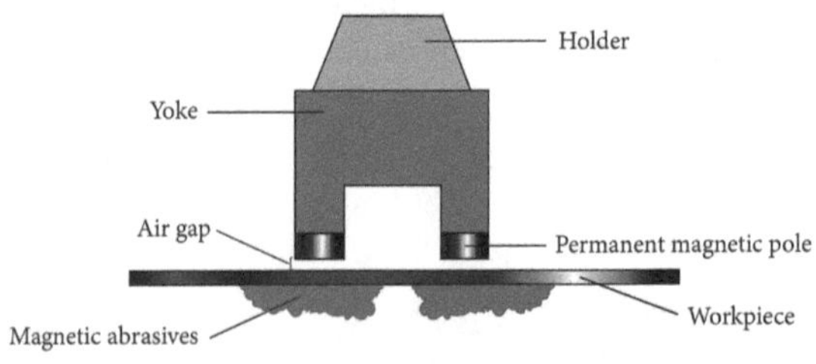

FIGURE 3.5 Schematic of MAF process for finishing of aluminium sheet.

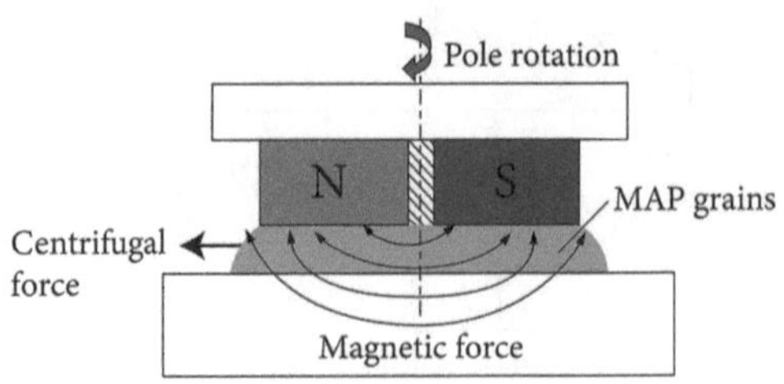

FIGURE 3.6 Mechanics of material removal in MAF flat surfaces of aluminium.

However, the effectiveness of higher rotation speeds diminished at extremely high rates due to the compensatory influence of reduced finishing pressure. The researchers also examined the effect of powder volume on the finishing process. They found that while increasing the powder volume had some positive effects, these effects were not significant and had limited impact on the overall finishing process.

After conducting their experiments, the researchers identified the ideal parameter values that led to significant enhancement in surface finish. The key parameters determined were as follows: the number of cycles was set to 6, the working gap was maintained at 1.5 mm, the rotational speed was set at 300 rpm, the weight of powder used was 2 g, and the process conditions applied were those specified in this study. They recommended a feed rate ranging from 20 mm/min to achieve desirable results. Ultimately, employing these parameter settings in the experiments resulted in a reduction of surface roughness (ΔRa) by 0.267 μm compared to its initial value.

3.4.4.2 Magnetorheological finishing (MRF)

MRF is a precise finishing process that utilizes a magnetic field to assist in achieving high surface accuracy and low surface roughness. It was developed and brought to market by QED Technologies Inc [20]. MRF can be used on a wide variety of materials, such as optical glasses and hard crystals. In MRF, the polishing fluid, known as the magnetorheological polishing (MRP) fluid, possesses unique rheological properties that can be controlled by adjusting the magnetizing current in the electromagnet coils. The carbonyl iron particles (CIPs) and ultra-fine abrasives are suspended in a carrier fluid in the MRP fluid. When a magnetic field is introduced into or removed from this fluid, reversible changes in its rheological properties occur. When a magnetic field is present, the CIPs acquire a magnetic dipole moment proportional to the strength of the field. They align themselves in chain-like structures in the direction of the field, creating interconnected columnar formations. Non-MAPs are inserted within or between these columnar structures. The rheological properties of the MRP fluid, along with the bonding strength between the abrasive particles and the CIPs in the presence of the magnetic field, are crucial for the effectiveness of the MRF process.

MRF offers a significant benefit through its exceptional precision and surface quality. It excels at achieving remarkably smooth surfaces with roughness at a sub-nanometre scale. This makes MRF particularly valuable in industries such as optics, electronics, and other advanced fields where precise surface finishing is of utmost importance. The versatility of MRF allows it to polish a diverse array of materials, ceramics, glass, and metals. It is a highly automated process that can be continuously monitored and controlled in real time, resulting in an efficient and consistent manufacturing method. As a result, MRF has emerged as a vital technology in numerous industries that rely on high-precision surface finishing.

3.4.4.2.1 Magnetorheological fluid (MR fluid)

MRF processes rely on the MR effect demonstrated by CIPs and abrasive particles suspended in a non-magnetic carrier medium. The MR fluid and its composition have a significant impact on MRF performance. MR fluids, which were first discovered by Rabinow in 1948, are classified as smart fluids due to their ability to exhibit variations in rheological behaviour in response to an applied magnetic field [21].

MR fluids are made up of micron-sized magnetic particles floating in a visco-elastic base medium such as water, silicone oil, glycerol, or paraffin oil, as well as certain additives. These fluids display non-Newtonian behaviour in the absence of a magnetic field, specifically weak Bingham behaviour. However, upon the application of a magnetic field, the fluid's rheological properties change, becoming stiffer and requiring a significant shear force to induce flow. It is important to note that the maximum strength of an MR fluid is restricted by its magnetic saturation point.

The carrier liquid chosen is critical in establishing the temperature ranges in which MR fluids can be used efficiently. Although silicon oil is commonly employed as the carrier liquid, hydrocarbon oil offers certain advantages due to its lower viscosity, improved lubrication properties, and suitability for high shear-rate applications. Additionally, an MR fluid based on hydrocarbon oil exhibits a lower zero field viscosity compared to its silicone oil-based counterpart, with the former being approximately 0.6 times less viscous. Conversely, a water-based MR fluid presents the benefits of minimizing waste disposal issues and facilitating easy particle recycling from the material [22].

In concentrated MR fluids, the presence of remnant magnetization in the particles leads to undesired particle aggregation. This phenomenon contributes to the formation of stiff sediments that are challenging to redisperse. To address this issue and mitigate particle aggregation and settling, various approaches have been proposed in the literature. These include:

1. Including thixotropic agents such as carbon fibres and silica nanoparticles [23, 25].
2. Incorporating surfactants like oleic or stearic acid [22].
3. Introducing magnetic nanoparticles [24, 25].
4. Utilizing viscoplastic media as a continuous phase [26].
5. Utilizing water-in-oil emulsions as carrier liquids [27].

These proposed methods aim to enhance the dispersion and stability of MR fluids by reducing the adverse effects of particle aggregation and sedimentation. By implementing these strategies, the performance and usability of MR fluids can be improved in various applications.

In a recent research study Khan and Jha [28] conducted experiments involving copper, a well-known diamagnetic material. Diamagnetic materials, when positioned in proximity to a permanent magnet, experience the induction of an equivalent magnetic field within them. However, it is essential to notice that the magnetic flux density within the diamagnetic material is lower than the magnetic flux density outside the diamagnetic material. As a result of the magnetization process in the diamagnetic material, the magnet and the induced magnetic field exert opposing forces on each other. This opposition causes the diamagnetic material to exhibit repulsion from the magnetic field, similar to the repulsion observed between two magnets when their magnetic flux densities oppose each other. It is essential to emphasize that the repelling force encountered by diamagnetic materials is exceedingly small in magnitude.

They looked into the magnetically controlled stiffened ball of an MR fluid that is used in the ball-end MR finishing process. The workpiece material is copper. In ball-end MR finishing, regardless of the type of base fluid, copper reacts with oxygen and the surface turns black and brownish in colour due to oxide formation. For the MR finishing of copper, BTA (benzotriazole)-mixed MRP fluid was created as a solution to this issue. To improve the finishing effect on copper, this synthetic fluid uses larger-particle electrolytic iron powder (rather than CIPs). When compared to CIPs, larger magnetizable particles (electrolytic iron powder) perform better in terms of scratch removal. According to the parametric analysis of the MRP fluid, increasing the abrasive concentration up to a certain limit increases the percent change in surface roughness (%ΔRa), but adding more abrasive causes the MRP fluid's magnetic permeability to decrease, which causes the percent change in surface roughness to decrease; %ΔRa increases as the concentration of electrolytic iron powder increases, but after a certain point (roughly 23 vol%), the results stabilize because the fluid is saturated in a magnetic field. According to an optimization study, the optimum concentrations of abrasive and EIP (electrolytic iron powder) for finishing copper are 14 vol% and 23 vol%, respectively. In 30 minutes of finishing, the surface roughness was decreased from 65.90 to 38 nm with a 42.34% improvement using the optimum polishing fluid composition.

A research study was carried out by Kansal et al. [2] on finishing of diamagnetic material using permanent magnets tool. In this study, the MR polishing fluid consists of 23% SiC abrasives, with particle sizes of 21 μm and 19 μm (two separate fluids were prepared). Additionally, it contained 17% CIPs with sizes of 37 μm, and a base fluid comprising 20% grease and 80% paraffin oil by weight, accounting for the remaining 60% of the composition. The selection of these compositions was based on the outcomes of preliminary experiments. Initially, SiC abrasive particles of 21 μm were employed for finishing the workpiece surface. Subsequently, SiC particles with a size of 19 μm were used for the final polishing of the surface of workpiece. For the experimentation process, a workpiece of a copper alloy with dimensions of $50 \times 40 \times 5\ mm^3$ was chosen. The finishing tool was rotated over the surface of the workpiece of copper alloy to achieve the desired surface finish.

This study investigates the effect of surface finishing on a workpiece of the copper alloy, focusing on the surface roughness parameters. Specifically, the roughness parameters Ra, Rq, and Rz were analysed before and after the finishing process. In the initial state, the workpiece exhibited surface roughness values of 273.6 nm, 347.2 nm, and 1769 nm for Ra, Rq, and Rz, respectively. During the finishing process, which lasted for 7.5 minutes, a notable improvement in surface roughness was observed. The roughness parameters decreased to 28.8 nm, 38.1 nm, and 269 nm for Ra, Rq, and Rz, respectively. This reduction in roughness of surface indicates the efficacy of the finishing technique in enhancing the workpiece's surface quality. Furthermore, an additional improvement in surface roughness was achieved by reducing the size of the abrasives employed in the finishing process. By decreasing the abrasive size from 21 to 19 μm, a further decrease in roughness of surface from 48.6 to 28.8 nm was attained. This finding highlights the importance of abrasive size selection in achieving desired surface finishes. This process can nano-finish diamagnetic materials like copper, which is extremely valuable in industries such as aerospace, dentistry, lasers, etc., for metallic mirrors.

Alam et al. [3] explored that, in the traditional finishing processes, it is hard to finish copper to the order of nanometres with uncontrolled natural forces. Also, copper's diamagnetic property limits the use of several complex magnetic-assisted finishing methods.

In their research, a novel technique for polishing nonplanar copper mirrors using MRP fluid was investigated. An effective slotted tip and solid rotating core electromagnet-based finishing tool were developed and mounted on a five-axis CNC system. When polishing diamagnetic copper workpieces, the solid spindle core and numerous slots at the tip increase the magnetic flux density and effectively grip the polishing fluid. In-process control of the finishing forces acting on the copper workpiece is also made easier by the electromagnet-based finishing tool. Using a dynamometer, forces acting on the workpiece during the finishing process were recorded and analysed to better understand the physics of the material removal mechanism. Using the novel tool, mirror-like reflective surfaces were created on two nonplanar copper workpieces (convex and concave samples). Surface roughness of the two samples was decreased in two finishing cycles with varying magnetizing current, from initial values of 120.1 and 129.2 nm to final values of 40.3 and 43.2 nm for convex and concave samples, respectively.

Hashmi et al. [29] developed a magnetic polishing tool with a 0.5-tesla permanent neodymium magnet (Nd2Fe14B) that is extremely powerful. For the purpose of finishing workpieces, this polishing tool was put together with the vertical milling machine. In this study, a developed MRP setup was used to conduct experiments on magnetic materials (steel material) and non-ferromagnetic (copper) content.

They looked into how various abrasives affected the magnetorheological finishing process parametrically. They assessed the impact of abrasives and magnetic particle concentration on the surface roughness of ferroelectric (copper) and non-ferroelectric (stainless steel) materials. For non-ferromagnetic (copper) material, the surface roughness value decreased from its starting value of 800 to 30 nm. The value was from 50 to 1300 nm for the magnetic material (stainless steel).

Ghosh et al. [30] observed that, in the electronics, space, and optical industries, a highly polished copper surface without any observable residual stress is in high demand. Due to its extremely poor hardness, copper cannot be properly finished using conventional finishing techniques.

The oxygen-free high conductivity (OFHC) copper's surface roughness is achieved in the current work using a wheel-based MRF procedure. The working gap, feed rate, and wheel speed are chosen as the primary process control parameters. As the response parameters, areal surface roughness (Sa), normal force, tangential force, and residual stress are studied. Because residual stress can change the completed surface's optical properties, measuring it is extremely important. Using the X-ray diffraction (XRD) technique, the residual stresses and phases of all the finished surfaces are analysed. To forecast the surface roughness of surfaces processed by MRF, a theoretical model is put forward. It has been found that by using MRF with a higher working gap, lower wheel speed, and feed rate, a nanolevel surface roughness with lower residual stress can be achieved. They reported that the OFHC copper surface had a minimum surface roughness (Sa) of 15.5 nm and a compressive residual stress of 6.9 MPa.

A new MR technique was envisaged by Kansal et al. [2] for the precise surface finishing of materials with diamagnetic properties. The diamagnetic copper alloy workpiece was polished using MRP fluid along with two cylindrical permanent magnets. Rotation over the copper alloy workpiece surface while using a cylindrical permanent magnet tool with MRP fluid at its tip surface finishes the copper alloy workpiece. Models and simulations have been done for the diamagnetic copper workpiece, MRP fluid at the tip surface, and finishing tool for permanent magnets. The working gap's magnetic flux density distribution is measured and examined. On a copper alloy workpiece, experiments are conducted, and the lowest Ra value of 28.8 nm from its initial value of 273.6 nm is obtained in 7.5 minutes of finishing time. Atomic force microscopy (AFM) and scanning electron microscopy (SEM) are used to analyse the surface characteristics of polished and unpolished workpieces. The results of the experimentation and finite element analysis provide assurance that the innovative design of the MRF tool utilizing permanent magnets is able to nano-finish diamagnetic materials like copper alloy and other similar materials.

3.4.5 Laser finishing

The working principle of laser finishing is based on the utilization of a high-energy laser beam to induce surface modifications in a material. By focusing the laser beam onto the material's surface, the energy imparted causes a significant increase in temperature, resulting in melting, vapourization, or ablation of the material. The laser beam's intensity can be precisely controlled, enabling a wide range of material removal and modification capabilities. The laser beam can be focused to a small spot size, allowing for high precision and the creation of intricate designs and patterns.

Distinct laser finishing techniques employ different laser settings, including power density, pulse duration, and wavelength. For instance, laser engraving typically employs high power density and short pulse duration to remove material and achieve deep engraving. Conversely, laser marking commonly employs lower power density and longer pulse duration to induce surface colour change or oxidation reactions without material removal.

The laser finishing process is influenced by material properties, such as thermal conductivity, melting point, and reflectivity, which necessitate careful selection and adjustment of laser settings and parameters according to the specific material and desired outcome.

The typical laser finishing process involves the following steps [31 :

Material preparation: The material to be processed undergoes cleaning, drying, and preparation procedures specifically tailored for laser finishing.

Design preparation: Specialized software or CAD tools are utilized to create the desired design or pattern to be engraved or etched onto the material.

Laser setup: The laser machine is calibrated and configured to match the requirements of the specific material and design under consideration.

Laser processing: The laser is employed to remove material, mark, etch, or engrave the material's surface as per the desired design. The laser energy can be adjusted to regulate the depth and width of engraved lines or patterns.

Quality control: The finished product undergoes inspection to ensure it conforms to the desired specifications and meets established quality standards.

Laser finishing exhibits versatility by being applicable to a wide range of materials, including metals, glass, plastics, and ceramics. It is a precise and versatile process capable of creating intricate and detailed designs on the material's surface. Furthermore, laser finishing is a non-contact procedure, eliminating physical contact between the material and the laser, thereby reducing the risk of damage or deformation.

The laser surface finishing of 3D printed components created by material extrusion was studied by Lambiase et al. [32]. The surface of components made of polylactic acid (PLA) generated by material extrusion (ME) was treated using a 30W CO_2 laser source operating in continuous wave (CW) mode. The mechanism of surface finishing was examined, and the effect of the laser treatment parameters on the surface morphology (scanning speed, energy density, and hatch distance) was established. Roughness measurements and optical microscopy were used to describe the surface of the treated samples. The outcomes showed that surface roughness could be greatly improved by laser finishing (more than ten times). However, interaction mechanisms incorporate surface ablation. To ascertain how the process parameters affected the surface properties, analysis of variance (ANOVA) was carried out. To find the ideal process conditions, the response surface method (RSM) and multi-response optimization (MRO) were also used. Thus, process parameters that reduced surface recession (0.1 mm) and increased surface roughness (from Ra = 16 μm for untreated surface to Ra = 0.3 μm under ideal process conditions) were identified.

In order to use dual-beam lasers to quickly polish steel free-form surfaces with a smooth surface finish, Zhou et al. [33] investigated an efficient "melting peak for filling with valley (MPFV)" technique. The dual-beam comprises a top-hat profiled pulsed laser beam and a CW laser beam. With high scanning speed of 800 mm/s, a CW laser polished an initial rough surface down to Sa (= 426 nm), while a pulsed laser played a significant role in applying the MPFV method to polish the surface down to Sa (= 142 nm), closely following the CW laser in the trajectory of a zigzag-square wave. A semisphere (about 60 mm in diameter) was polished in only about 35.5 seconds, which led to a roughness decrease from the initial state value of Sa (= 877 nm) to the post-polished value of Sa (= 142 nm) and a smoother surface finish. The effectiveness of dual-beam laser polishing is above 2890 cm^2/H, which is more than 100 times that of manual polishing. This study's key finding is that the zigzag-square trajectory makes the pulsed laser that follows the CW laser implement the MPFV method more effectively and efficiently than the standard process.

REFERENCES

1. Mahalik, N.P., Jha, S., & Jain, V.K. (2006). Nanofinishing techniques. *Micromanufacturing and Nanotechnology*, 171–195.
2. Kansal, H., Singh, A. K., & Grover, V. (2018). Magnetorheological nano-finishing of diamagnetic material using permanent magnets tool. *Precision Engineering*, *51*, 30–39.
3. Alam, Z., Khan, D. A., & Jha, S. (2019). MR fluid-based novel finishing process for nonplanar copper mirrors. *The International Journal of Advanced Manufacturing Technology*, *101*, 995–1006.

4. Givi, M., Fadaei Tehrani, A., & Mohammadi, A. (2012). Polishing of the aluminum sheets with magnetic abrasive finishing method. *The International Journal of Advanced Manufacturing Technology*, *61*, 989–998.
5. Liu, D., Zhang, Z., Feng, J., Yu, Z., Meng, F., Shi, C., ... & Liu, W. (2023). Environment-friendly chemical mechanical polishing for copper with atomic surface confirmed by transmission electron microscopy. *Colloids and Surfaces A: Physicochemical and Engineering Aspects*, *656*, 130500.
6. Kaushik, R. M., Bhandakkar, A. B., & Patro, T. U. (2014). Solution of emulsifiable oil and hydrogen peroxide for chemical–mechanical polishing of Ti alloy—A green approach. *Materials Letters*, *122*, 252–255.
7. Zhang, Z., Liao, L., Wang, X., Xie, W., & Guo, D. (2020). Development of a novel chemical mechanical polishing slurry and its polishing mechanisms on a nickel alloy. *Applied Surface Science*, *506*, 144670.
8. Gupta, A., Saini, A., Khatri, N., & Juyal, A. (2022). Review of single-point diamond turning process on IR optical materials. *Materials Today: Proceedings*, *69*, 435–440.
9. Li, L. H., Yu, N. H., Chan, C. Y., & Lee, W. B. (2018). Al6061 surface roughness and optical reflectance when machined by single point diamond turning at a low feed rate. *Plos One*, *13*(4), e0195083.
10 Zhang, G., Han, J., Chen, Y., Xiong, J., Wang, J., & Ran, J. (2022). Generation mechanism and dual-dynamics simulation of surface patterns in single-point diamond turning of single-crystal copper. *Journal of Manufacturing Processes*, *75*, 1023–1038.
11. Zong, W. J., Huang, Y. H., Zhang, Y. L., & Sun, T. (2014). Conservation law of surface roughness in single point diamond turning. *International Journal of Machine Tools and Manufacture*, *84*, 58–63.
12. Zhao, D., & Lu, X. (2013). Chemical mechanical polishing: theory and experiment. *Friction*, *1*, 306–326.
13 Joshi, M., More, S., Singh, R. K., Joshi, S. S., Balasubramaniam, R., & Suri, V. K. (2012). Experimental characterization of hydrodynamic nanopolishing of flat steel plates. *Precision Engineering*, *36*(3), 424–434.
14. Lebedeva, O., Kultin, D., Zakharov, A., & Kustov, L. (2022). Advances in application of ionic liquids: fabrication of surface nanoscale oxide structures by anodization of metals and alloys. *Surfaces and Interfaces*, *34*, 102345.
15. Sun, X., & Zou, Y. (2018). Study on electrolytic magnetic abrasive finishing for finishing stainless steel SUS304 plane with a special compound machining tool. *Journal of Manufacturing and Materials Processing*, *2*(3), 41.
16. Kawanaka, T., & Kunieda, M. (2015). Mirror-like finishing by electrolyte jet machining. *CIRP Annals*, *64*(1), 237–240.
17. Maehata, H., Kamada, H., & Yamamoto, M. (1987). Electrolytic-abrasive mirror finishing. *Precision Engineering*, *9*(1), 31–43.
18. Wang, K., Yan, Y., Zhou, P., Zhang, C., Kang, R., & Guo, D. (2020). Preparation of flat and smooth copper surface by jet electrochemical machining and electrochemical polishing. *Journal of The Electrochemical Society*, *167*(16), 163501.
19. Qian, C., Fan, Z., Tian, Y., Liu, Y., Han, J., & Wang, J. (2021). A review on magnetic abrasive finishing. *The International Journal of Advanced Manufacturing Technology*, *112*, 619–634.
20. Saraswathamma, K. (2014). Magnetorheological finishing: a review. *International Journal of Current Engineering and Technology, Special Issue-2*. http://Dx.Doi.Org/10.14741/Ijcet/Spl.2.2014.30
21. Rabinow, J. (1948). The magnetic fluid clutch. *Electrical Engineering*, *67*(12), 1167–1167.
22. Phulé, P. P., & Ginder, J. M. (1999). Synthesis and properties of novel magnetorheological fluids having improved stability and redispersibility. *International Journal of Modern Physics B*, *13*(14n16), 2019–2027.

23. Bossis, G., Volkova, O., Lacis, S., & Meunier, A. (2002). Magnetorheology: fluids, structures and rheology. *Ferrofluids: Magnetically Controllable Fluids and Their Applications*, 202–230. Lecture Notes in Physics, Volume 594, Springer.
24. Chin, B. D., Park, J. H., Kwon, M. H., & Park, O. O. (2001). Rheological properties and dispersion stability of magnetorheological (MR) suspensions. *Rheologica Acta*, *40*, 211–219.
25. López-López, M. T., De Vicente, J., González-Caballero, F., & Durán, J. D. G. (2005). Stability of magnetizable colloidal suspensions by addition of oleic acid and silica nanoparticles. *Colloids and Surfaces A: Physicochemical and Engineering Aspects*, *264*(1–3), 75–81.
26. Rankin, P. J., Horvath, A. T., & Klingenberg, D. J. (1999). Magnetorheology in viscoplastic media. *Rheologica Acta*, *38*(5), 471–477.
27. Park, J. H., Chin, B. D., & Park, O. O. (2001). Rheological properties and stabilization of magnetorheological fluids in a water-in-oil emulsion. *Journal of Colloid and Interface Science*, *240*(1), 349–354.
28. Khan, D. A., & Jha, S. (2018). Synthesis of polishing fluid and novel approach for nanofinishing of copper using ball-end magnetorheological finishing process. *Materials and Manufacturing Processes*, *33*(11), 1150–1159.
29. Hashmi, A. W., Mali, H. S., Meena, A., Khilji, I. A., & Chilakamarry, C. R. (2022). Experimental investigation on magnetorheological finishing process parameters. *Materials Today: Proceedings*, *48*, 1892–1898.
30. Ghosh, G., Sidpara, A., & Bandyopadhyay, P. P. (2021). Experimental and theoretical investigation into surface roughness and residual stress in magnetorheological finishing of OFHC copper. *Journal of Materials Processing Technology*, *288*, 116899.
31. Krishnan, A., & Fang, F. (2019). Review on mechanism and process of surface polishing using lasers. *Frontiers of Mechanical Engineering*, *14*, 299–319.
32. Lambiase, F., Genna, S., & Leone, C. (2020). Laser finishing of 3D printed parts produced by material extrusion. *Optics and Lasers in Engineering*, *124*, 105801.
33. Zhou, Y., Zhao, Z., Zhang, W., Xiao, H., & Xu, X. (2019). Experiment study of rapid laser polishing of freeform steel surface by dual-beam. *Coatings*, *9*(5), 324.

4 Exploring the frontiers of nanofinishing for biomedical implant applications in healthcare

Vidyapati Kumar, Ankita Mistri, Rajan Kumar Verma, Varnit Jain, and Manojit Ghosh

4.1 INTRODUCTION

Nanofinishing uses nanotechnology in surface modification processes for diverse materials, including bioimplants. It entails modifying and regulating materials at the nanoscale level to improve their characteristics and capabilities. Nanofinishing methods may be utilized to enhance the surface properties of bio-implants, such as roughness, wettability, and biocompatibility. Bio-implants are biological substitutes to repair, maintain, or change wounded tissues and organ function [1]. They play an essential role in orthopaedic and dental applications, providing solutions for various medical issues. When creating bio-implants, biocompatibility is crucial. It refers to the implant material's compatibility with the biological environment, which ensures the implant can operate appropriately without generating undesirable responses in the body. The study of nanofinishing has brought about revolutionary improvements in biomedical implant applications, revolutionizing the manufacturing process and giving enormous possibilities for improving surface characteristics critical to obtaining optimum functional performance [2]. The primary goal in biomedical implant applications is to achieve particular surface finishes on implanted components. These surface qualities guarantee the implants' biocompatibility, longevity, and general performance. Nanofinishing techniques have emerged as critical tools for attaining these same surface finishes while protecting the integrity of the implant surfaces [3]. This study focuses on understanding the vital role that finishing procedures play in achieving specified surface qualities for optimum functional performance. These activities, which occur at the end of the manufacturing process, reduce the surface roughness, eradicate flaws, and improve the integrity of the implant surfaces [4]. Nanofinishing procedures give the precision and control required to overcome the constraints of older approaches, enabling us to unleash the full potential of implant surfaces and deliver more excellent clinical results.

DOI: 10.1201/9781003496298-4

The need for specialized techniques is undeniable when working with complicated free-form surfaces, which involve flexible finishing equipment capable of attaining nanometre-level surface finishes while assuring the absence of any surface or sub-surface flaws. Notably, this need occurs in various applications, such as the production of human implants, as shown in Figure 4.1, where free-form surfaces are prominent [5].

The use of nanofinishing technologies in biomedical implants brings new concerns and potential. Complex implant surfaces, such as knee joints, hip joints, and turbine blades, provide unique problems requiring nanometre-level surface finishes with no surface or sub-surface flaws [6]. Engineers have investigated the possibilities of computer numerical control (CNC) and robotics to offer exact relative motion between the workpiece and the finishing tool, allowing the creation of specialized nanofinishing procedures adapted to free-form surfaces.

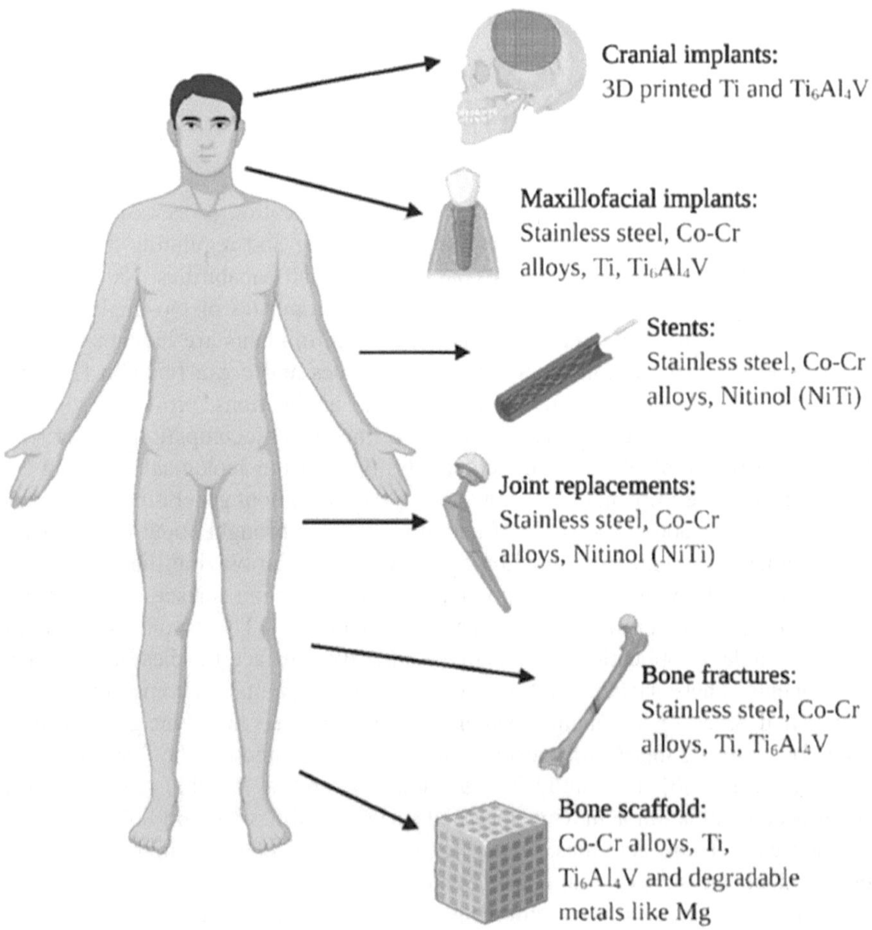

FIGURE 4.1 Diverse metallic implants utilized in the human body [5].

4.2 BIOMEDICAL IMPLANTS AND BIOCOMPATIBILITY

Various factors influence medical implants' biocompatibility, as shown in Figure 4.2. Material selection involves choosing biocompatible materials for the implant. Surface characteristics involve the properties of the implant surface, including roughness and topography. Design considerations focus on factors like shape and biomechanical compatibility. Patient factors encompass individual health and specific patient characteristics. Corrosion resistance aims to prevent material degradation and corrosion. Mechanical compatibility ensures that the implant's mechanical properties align with biomechanical requirements. Surface morphology considers the roughness and topography of the implant surface. Chemical stability involves maintaining stability and compatibility with physiological conditions. Immunological response minimizes immune reactions and promotes favourable tissue responses. Sterility and infection control emphasize the importance of sterile manufacturing and infection prevention. Degradation properties focus on controlling the degradation rate of biodegradable implants. Additional factors include patient-specific conditions, healing capacity, and

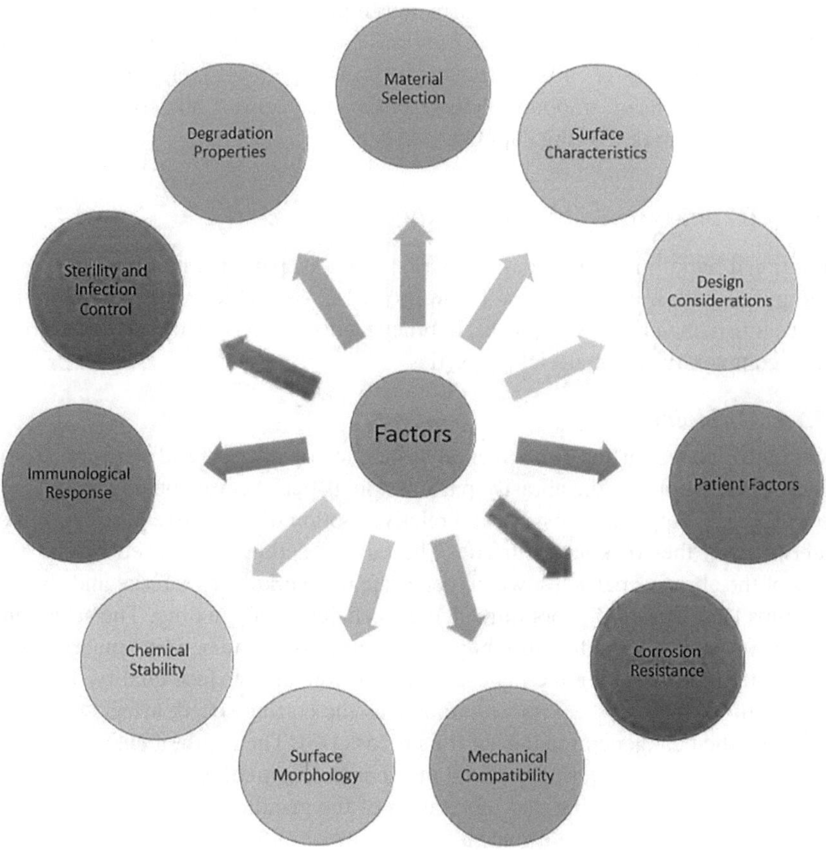

FIGURE 4.2 Factors affecting the biocompatibility of medical implants.

metabolic considerations. Considering these factors is crucial for ensuring the biocompatibility and success of medical implants.

Implant integration into the human body depends on structural design and functioning, the materials employed, and their interaction with biological processes. Understanding the qualities and features of various materials is critical in this context for creating successful and biocompatible implants.

Table 4.1 summarizes the most often used materials in medical implants and their essential features and uses. These materials were carefully selected for their unique mix of strength, biocompatibility, corrosion resistance, and other vital characteristics. Healthcare providers may improve patient outcomes and encourage effective implant integration by adapting material choices to individual implant needs.

4.3 CLASSIFICATION OF FINISHING METHODS FOR BIOMEDICAL IMPLANTS

The surface properties of biomedical implants play a critical role in determining their biocompatibility and overall performance in healthcare applications. Nanofinishing techniques offer a precise way to control and improve these surface properties at the nanoscale level. This section discusses various nanofinishing techniques, as depicted in Figure 4.3, such as chemical-mechanical polishing (CMP), electrochemical polishing (ECP), ion beam smoothing (IBS), ultrasonic peening, abrasive flow machining, vapour polishing, buffing, lapping.

4.3.1 Traditional nanofinishing processes

Traditional nanofinishing processes (TFPs), such as grinding, honing, lapping, and buffing, fall within this category. These processes can achieve the nanometre-level surface finish on various workpieces, but they have certain constraints related to size, shape, complexity, and material properties.

4.3.1.1 Grinding

Grinding operates on the fundamental principles of friction and abrasion. As the grinding wheel rotates, the abrasive particles on its periphery come into contact with the workpiece. Applying pressure and relative motion, the abrasive particles remove material from the workpiece surface. The cutting action is achieved by the sharp edges of the abrasive particles, which penetrate the workpiece material and shear off tiny chips [13]. Several factors influence the efficiency of grinding. The selection of the grinding wheel, including its composition, grain size, and bonding material, plays a vital role in determining the material removal rate and surface quality. The speed at which the grinding wheel rotates, known as the cutting speed, affects the cutting action and the heat generated during the process [13]. The depth of cut, which determines the amount of material removed in each pass, is another crucial parameter.

Furthermore, the cooling and lubrication of the grinding zone are essential both for thermal control and for preventing excessive wear of the grinding wheel. Coolants and lubricants dissipate the heat generated, facilitate chip evacuation, and improve

TABLE 4.1
Materials used in medical implants and their properties

Material	Properties	Applications
Titanium [7]	High strength, low density, excellent corrosion resistance, biocompatible, good osseointegration	Orthopaedic implants (hip, knee, spinal), dental implants, cardiovascular implants (stents, pacemakers), surgical instruments
Stainless steel [7]	High mechanical strength and durability, good corrosion resistance, biocompatible, MRI compatibility	Orthopaedic implants (fracture fixation devices), dental implants, cardiovascular implants (stents, valves), surgical instruments
Cobalt-chromium alloy [3]	High strength, wear resistance, excellent corrosion resistance, biocompatible, low metal ion release	Orthopaedic implants (knee, hip, dental), cardiovascular implants (stents, heart valves), prosthetics, bone plates and screws
Polymers [7]	Lightweight, flexible, biocompatible, tailorable mechanical properties, low friction and wear	Soft tissue implants (breast, facial), drug delivery systems, orthopaedic implants (suture anchors), tissue engineering scaffolds
Ceramic materials [8]	High compressive strength, bioinert or bioactive properties, excellent biocompatibility, wear resistance, good osseointegration	Dental implants, orthopaedic implants (hip, knee, spinal), spinal implants, maxillofacial implants, bone graft substitutes
Composite materials [9]	Tailorable combination of properties (strength, flexibility, biocompatibility), enhanced mechanical performance, variable degradation rates, and versatile design options	Dental implants, orthopaedic implants (bone fixation devices), tissue engineering constructs, custom implants, craniofacial implants
Nitinol [10]	Shape memory effect, superelasticity, corrosion resistance, biocompatible	Orthopaedic implants (stents, bone fixation devices), dental implants
PEEK [7]	High strength, radiolucent, biocompatible, low wear, chemical resistance	Spinal implants, orthopaedic implants, dental implants
Hydrogels [4]	High water content, excellent biocompatibility, soft and flexible	Soft tissue implants, drug delivery systems, tissue engineering scaffolds
Magnesium alloys [11]	Biodegradable, biocompatible, mechanical properties similar to bone	Bone fixation devices, cardiovascular implants, orthopaedic implants
Bioactive glasses [12]	Bioactive properties, good osseointegration, biocompatible	Dental implants, bone graft substitutes, tissue engineering scaffolds
Silk [12]	Biocompatible, biodegradable, mechanical properties similar to soft tissues	Soft tissue implants, wound healing scaffolds, drug delivery systems

FIGURE 4.3 Various nanofinishing techniques.

surface finish. The choice of coolant or lubricant depends on the specific requirements of the grinding operation and the workpiece material.

4.3.1.2 Honing

Honing relies on the principles of controlled abrasion to refine surfaces with precision. During the honing process, an abrasive stone, also known as a hone, is pressed against the workpiece surface and moved along a defined path [13]. The honing stones consist of abrasive particles embedded in a bonding material, which provides stability and abrasion resistance. The honing action is achieved by interacting with the abrasive particles and the workpiece material. The honing stone removes material through cutting and ploughing as it moves across the surface [13]. The cutting occurs when the sharp edges of the abrasive particles shear off thin layers of material, while the ploughing step involves material displacement through pressure and friction. Various factors influence the performance of honing. The rotational speed of the honing tool, the stroke length, and the applied pressure determine the material removal rate and surface quality [13]. The selection of appropriate honing stones with the desired grit size and bonding material is critical for achieving the desired precision and surface finish. Additionally, using lubricants or honing oils helps maintain a consistent cutting action, reducing friction and dissipating heat.

4.3.1.3 Lapping

Lapping is a nanofinishing process that utilizes a flat abrasive material known as a lap to grind and polish the surface of a workpiece. It is beneficial for achieving high surface finish and flatness levels required in applications such as optical lenses, semiconductor wafers, precision mechanical parts, and biomedical implants that require a smooth and polished surface to prevent implant rejection [13]. Soft materials like cast iron or copper are used as a lap and coated with an abrasive paste or slurry to perform lapping. The workpiece is then placed on the lap and moved in a circular or oscillating motion while pressure is applied. As the lap rotates, it removes material from the workpiece, smoothing and flattening the surface. The size and type of abrasive particles used in the slurry can be adjusted to achieve a wide range of surface finishes, from rough to mirror-like. Despite its benefits, lapping can be time-consuming and expensive, particularly for large workpieces or materials that are difficult to lap, such as ceramics and hard metals [13].

Moreover, achieving consistent and accurate results requires skilled operators. However, lapping remains a valuable nanofinishing technique for producing high-quality surfaces with precise flatness and surface finish for various applications. In the present context, lapping is considered one of the most effective nanofinishing processes for achieving high surface finish and flatness levels in multiple industries, such as optical manufacturing, semiconductor fabrication, and precision mechanical parts manufacturing [13]. It is also relevant in biomedical implants, where a smooth and polished surface is essential for proper functioning and to prevent implant rejection. Despite its limitations, lapping remains a valuable nanofinishing technique for producing high-precision surfaces.

4.3.1.4 Buffing

Buffing is a surface finishing process involving a buffing wheel or pad to remove imperfections from a material's surface. The method uses a combination of friction and pressure to smooth and shine the surface. The buffing wheel is typically made of cotton, linen, or leather and can be impregnated with abrasives such as diamond or aluminium oxide for more aggressive material removal [13]. Buffing is often used in the final stages of the manufacturing process to enhance the appearance of the product.

In the context of nanofinishing for biomedical implant applications in healthcare, buffing can be used to smooth and polish the surface of the implant to a high degree of smoothness and shine. This can reduce the friction between the implant and surrounding tissue, which can help to minimize tissue damage, inflammation, and other adverse effects. Additionally, a highly polished surface can reduce the risk of bacterial adhesion and biofilm formation, leading to implant failure and infection [13]. Buffing can be combined with other hybrid nanofinishing techniques, such as CMP or ECP, to achieve a high level of surface finish and biocompatibility for medical implants.

These traditional nanofinishing processes play a significant role in achieving desired surface properties for biomedical implants. They offer precision and control in improving surface smoothness, dimensional accuracy, and geometrical form [13]. By employing these processes, the surface finish can be enhanced to meet the stringent requirements of medical implants, ensuring their biocompatibility and overall performance in healthcare applications.

4.3.2 Hybrid nanofinishing techniques

4.3.2.1 Chemical-mechanical polishing

CMP is a nanofinishing process used to achieve a high degree of surface smoothness on various materials. The CMP process involves using a chemical slurry consisting of abrasive particles suspended in a chemical solution and a polishing pad. The chemical slurry is applied to the material surface, and the polishing pad is pressed onto it, with controlled pressure and speed, to achieve the desired surface finish [14]. CMP is widely used in the biomedical implant industry because it produces a smooth, defect-free surface with excellent biocompatibility. CMP can be used on various materials, including metals, ceramics, and polymers, making it a versatile technique for medical implant manufacturing. The process can be used for different stages of implant fabrication, such as surface preparation, roughness reduction, and final polishing.

One of the main advantages of CMP is its ability to provide a high degree of surface uniformity across the entire surface of the material, including curved and irregular surfaces. This is achieved through the chemical action of the slurry, which chemically reacts with the material to remove any defects, along with the mechanical movement of the polishing pad, which mechanically removes any remaining imperfections [15]. Furthermore, CMP can produce surfaces with a wide range of roughness values, from sub-nanometre to micrometre scales, depending on the size and type of abrasive particles used and the chemical properties of the slurry. This flexibility allows for producing surfaces with tailored roughness values for specific applications and materials. However, one of the limitations of CMP is the potential for contamination of the surface due to the use of the chemical slurry. Careful control of the slurry composition and the polishing parameters is essential to avoid any unwanted chemical reactions or impurities on the surface.

Additionally, the polishing pad must be carefully selected to avoid scratching or marring the surface. In conclusion, CMP is a powerful nanofinishing technique widely used in the medical implant industry due to its ability to produce smooth, uniform, and biocompatible surfaces on various materials. The flexibility of the technique allows for the production of tailored surfaces with specific roughness values for particular applications. However, careful control of the process parameters is required to avoid surface contamination and ensure consistent results.

4.3.2.2 Electrochemical polishing

ECP is a surface finishing technique that involves the removal of material from the surface of a metal by anodic dissolution in an electrolyte solution. In ECP, the workpiece is immersed in an electrolyte solution while a current is applied between the workpiece and an electrode [16]. The anodic current causes the metal ions to dissolve from the surface, resulting in a polished surface. ECP has several advantages over other surface finishing techniques, including removing surface defects, increasing corrosion resistance, and improving the surface finish. Additionally, ECP can be performed on complex shapes and delicate parts, making it a valuable tool in manufacturing biomedical implants. ECP is commonly used for polishing metals such as titanium and its alloys, stainless steel, and cobalt-chromium alloys. The electrolyte solution used for ECP depends on the metal being polished and the desired surface finish. For example, phosphoric acid and its derivatives are often used for titanium alloys, while sulphuric acid is used for stainless steel [17]. One of the critical factors in ECP is controlling the current density and electrolyte flow rate to achieve the desired surface finish. Additives, such as surfactants and chelating agents, can also improve the surface finish and reduce the amount of material removal. Overall, ECP is a versatile surface finishing technique with many applications in biomedical implants due to its ability to produce high-quality, defect-free surfaces with improved biocompatibility and corrosion resistance [17].

4.3.3 Advanced nanofinishing techniques

4.3.3.1 Ion beam smoothing

IBS is a nanofinishing technique that uses a beam of ions to remove surface material and create a smoother surface finish. IBS is often used on metals, semiconductors, and other materials requiring precise surface smoothing, such as biomedical implants [18]. In IBS, a beam of charged ions is directed at the material surface, causing atoms to be ejected and the surface to be smoothed out. The process can be controlled with high precision, allowing for excellent adjustments to the surface topography. The ions used in IBS are typically argon or helium ions, which are accelerated to high speeds and focused onto the material surface. IBS can achieve surface roughness on the order of nanometres and can be used to create surfaces with various shapes, including concave and convex surfaces [19]. The significant advantage of IBS is its ability to smooth out surfaces without damaging the underlying material or changing its properties. This makes it a beneficial technique for creating biomedical implants, where biocompatibility and surface smoothness are critical for implant success.

4.3.3.2 Ultrasonic peening

Ultrasonic peening is a nanofinishing technique that uses high-frequency mechanical vibrations to create small amplitude oscillations on the implant's surface. These oscillations generate a localized stress field that induces plastic deformation, which helps to smooth out surface irregularities and improve the surface finish. The ultrasonic waves are delivered to the surface using a sonotrode or a tool that vibrates at 20 kHz or higher [20]. The tool is typically made of a hard material such as tungsten carbide or diamond. The peening is done in a liquid medium containing abrasive particles, which helps to enhance the polishing action. The process can be carried out in either a batch or a continuous mode [20]. Ultrasonic peening is particularly useful for finishing complex shapes and hard-to-reach areas. It has been shown to improve biomedical implants' surface roughness, hardness, and wear resistance, thereby enhancing their biocompatibility and longevity.

4.3.3.3 Abrasive flow machining

Abrasive flow machining is a non-traditional machining process that is used for the precision finishing of complex internal geometries of metallic, ceramic, and composite materials. The process involves using a semi-solid, viscoelastic abrasive media forced through the workpiece using a hydraulic or pneumatic pressure system [21]. As the abrasive media flows through the workpiece, it removes the unwanted material from the internal surfaces, resulting in an improved surface finish, reduced roughness, and improved dimensional accuracy. The process is beneficial for finishing complex shapes such as impellers, turbine blades, and nozzles. The advantages of abrasive flow machining over traditional finishing methods include improved accuracy, reduced tool wear, and improved surface finish [21]. The process can also be automated, making it ideal for mass production of high-precision components. Abrasive flow machining has been used for various biomedical implant applications, including improving the surface finish of orthopaedic implants, dental implants, and cardiovascular stents.

4.3.3.4 Vapour polishing

Vapour polishing is a nanofinishing technique that involves exposing the surface of a material to a vapour of a suitable solvent or etchant, which causes the surface to undergo controlled dissolution. The process is typically carried out in a vacuum chamber, where the material is heated to a temperature that is above its melting or sublimation point but below its boiling point to create a thin layer of liquid on the surface [22]. The vapour of the solvent or etchant is then introduced into the chamber, where it reacts with the liquid layer to remove the material from the surface through condensation. Vapour polishing can be used to achieve extremely smooth and uniform surface finishes on a wide range of materials, including metals, ceramics, and polymers. The technique is beneficial for polishing complex geometries and delicate structures, as it does not involve any physical contact with the surface, eliminating the risk of mechanical damage or distortion [22]. In biomedical implants, vapour polishing has been used to improve the surface smoothness and biocompatibility of materials such as titanium, stainless steel, and polyether ether ketone (PEEK), commonly used in implantable medical devices.

4.4 CLASSIFICATION OF SURFACE CHARACTERIZATION FOR BIOMEDICAL IMPLANTS

Surface characterization is a fundamental process that involves analysing physical and chemical properties of a material surface. It plays a critical role in comprehending and optimizing the properties of implant materials, making it an indispensable component of biomedical engineering. By meticulously controlling the surface properties of biomedical implants, such as roughness, chemical composition, and topography, it is possible to enhance their biocompatibility, corrosion resistance, wear

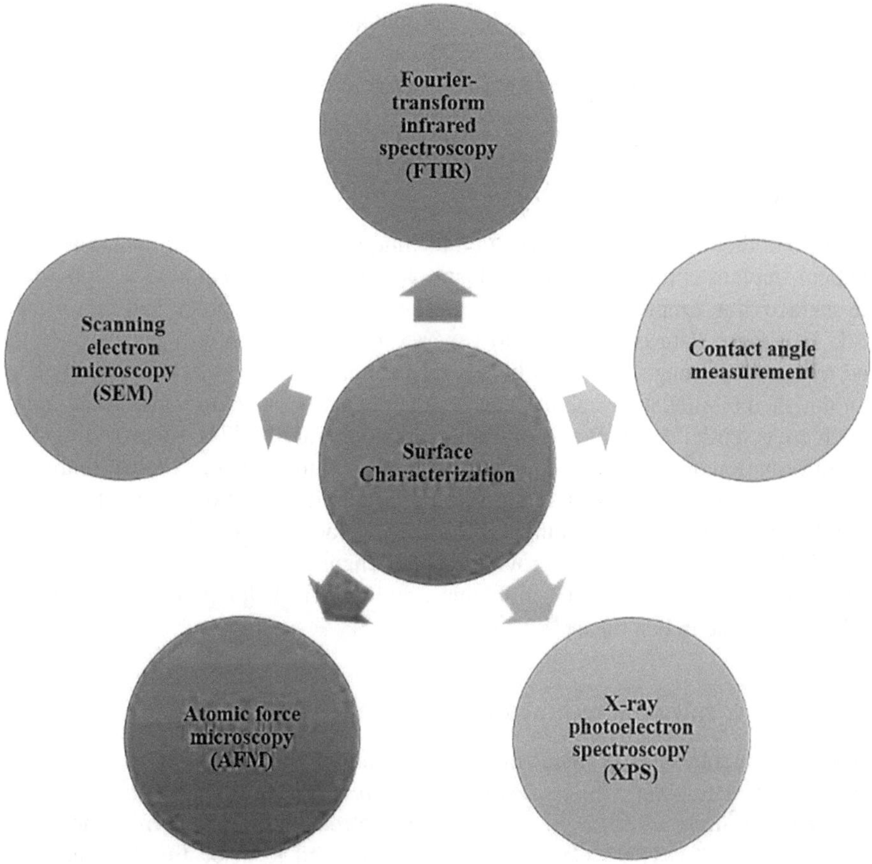

FIGURE 4.4 Various surface characterization techniques.

resistance, and other essential characteristics [5]. Figure 4.4 depicts various surface characterization techniques such as scanning electron microscopy (SEM), atomic force microscopy (AFM), X-ray photoelectron spectroscopy (XPS), contact angle measurement, and Fourier-transform infrared spectroscopy (FTIR).

4.4.1 Scanning electron microscopy

SEM is a powerful imaging technique that has significant relevance for healthcare, especially in nanofinishing for biomedical implant applications. SEM directs a focused beam of electrons at a sample surface, causing it to emit secondary electrons that can be detected and used to create high-resolution images of the surface morphology and topography [23]. The technique is beneficial for analysing the surface characteristics of nanofinished biomedical implants, such as their roughness and defects, which are critical determinants of biocompatibility [23]. By providing detailed information on the surface properties of implants, SEM can help researchers and manufacturers develop and optimize nanofinishing methods for improved implant biocompatibility and longevity. Combined with other surface characterization techniques like AFM and XPS, SEM can comprehensively analyse biomedical implants' biocompatibility and surface characteristics, facilitating their safe and effective use in healthcare.

4.4.2 Atomic force microscopy

AFM is another surface characterization technique critical in nanofinishing for biomedical implant applications in healthcare. AFM works by scanning a probe over the surface of a sample and measuring the force between the probe and the surface [24]. This force data is then used to generate high-resolution surface morphology and topography images. AFM is beneficial for analysing the surface properties of nanofinished biomedical implants, such as their roughness, texture, and topography, which are essential determinants of implant biocompatibility [24]. By providing precise information on the surface characteristics of implants, AFM can help researchers and manufacturers optimize nanofinishing methods for improved implant biocompatibility and longevity. Combined with other surface characterization techniques like SEM and XPS, AFM can provide a comprehensive analysis of the biocompatibility and surface characteristics of biomedical implants, helping to ensure their safe and effective use in healthcare.

4.4.3 X-ray photoelectron spectroscopy

XPS is a valuable surface characterization technique widely used in nanofinishing for biomedical implant applications in healthcare. XPS works by bombarding the surface of a sample with X-rays, which cause electrons to be ejected from the atoms in the sample [25]. These electrons are then analysed to determine the chemical composition of the surface. XPS is beneficial for studying the surface chemistry and elemental composition of nanofinished biomedical implants, which are essential determinants of implant biocompatibility [25]. By providing precise information on implants' surface chemistry and elemental composition, XPS can help researchers and manufacturers optimize nanofinishing methods for improved implant biocompatibility and longevity. In combination with other surface characterization techniques like SEM and AFM, XPS can provide a comprehensive analysis of the biocompatibility and surface characteristics of biomedical implants, helping to ensure their safe and effective use in healthcare.

4.4.4 Contact angle measurement

Contact angle measurement is a widely used technique for analysing biomedical implants' surface wettability and surface energy. It involves placing a droplet of liquid on the implant's surface and measuring the liquid's angle with the surface [26]. The contact angle measures the degree of wetting of the surface by the liquid, which in turn provides information about the surface energy and wettability of the implant [27]. A low contact angle indicates high surface wettability and low surface energy, while a high contact angle indicates low surface wettability and high surface energy. The contact angle measurement results can be used to optimize the surface properties of biomedical implants for specific applications, such as promoting cell adhesion or inhibiting bacterial colonization. Thus, contact angle measurement is an essential tool in nanofinishing for biomedical implant applications in healthcare, helping improve medical implants' performance and longevity.

4.4.5 Fourier-transform infrared spectroscopy

Fourier-transform infrared spectroscopy (FTIR) is a powerful analytical technique that can be used to analyse the chemical bonding and functional groups on the surface of implants. It works by shining infrared light on the implant's surface and measuring how much of the light is absorbed by the material [28]. Different chemical groups and functional bonds absorb infrared light at different frequencies, allowing FTIR to identify and quantify the types of chemical bonds and functional groups on the surface. This information is crucial for understanding the surface chemistry of implants and how they interact with biological tissues. By analysing the FTIR spectra of nanofinished implants, researchers can gain insights into the chemical composition of the surface and optimize nanofinishing parameters to achieve the desired chemical properties for improved biocompatibility [29]. FTIR is an essential tool in biomedical implant research, as it provides a non-destructive, quantitative method for analysing the chemical properties of implant surfaces.

4.5 CURRENT CHALLENGES IN NANOFINISHING FOR BIOMEDICAL IMPLANTS

As nanofinishing techniques advance, several challenges must be addressed to fully harness their potential in biomedical implant applications. This section explores the current challenges in nanofinishing for biomedical implants, their limitations, potential solutions, impact on healthcare, and future directions, as shown in Table 4.2. By overcoming these challenges, researchers and clinicians can pave the way for improved implant performance, enhanced biocompatibility, and personalized patient care.

4.6 EMERGING TRENDS FOR THE FUTURE OF NANOFINISHING IN BIOMEDICAL IMPLANT APPLICATIONS

In addition to addressing the current challenges in nanofinishing for biomedical implants, it is crucial to explore the emerging trends and future directions that hold great promise for advancing this field. These trends encompass the development of new nanofinishing methods and other surface modification techniques, and the impact of nanofinishing on implant longevity, wear and tear, and other crucial factors. By embracing these trends, researchers and industry professionals can further enhance the performance and effectiveness of biomedical implants, leading to improved patient outcomes and healthcare advancements.

TABLE 4.2
Critical issues in nanofinishing for biomedical implants

Challenge	Limitations	Solutions	Impact on healthcare	Future directions
Achieving consistent and reproducible results [30]	Lack of standardized protocols and quality control measures	Development of standardized protocols and quality control measures; advanced characterization techniques	Enhanced reliability and consistency of implant performance	Implementation of automated nanofinishing processes; utilization of machine learning algorithms for process optimization
Clinical translation and regulatory approval [30]	Stringent regulatory processes and safety considerations	Comprehensive evaluation framework; collaboration with regulatory bodies; optimization of manufacturing processes	Streamlined regulatory approval process; faster time to market for new implants	Integration of nanofinishing techniques into regulatory guidelines; establishment of dedicated regulatory pathways for nanofinished implants
Biological response and mechanisms [30]	Limited understanding of interactions at the nanoscale	Advanced characterization techniques; experimental studies; computational modelling and simulations	Enhanced biocompatibility and reduced risk of adverse reactions	In-depth exploration of nanoscale interactions through advanced imaging techniques and biomolecular studies
Material selection and compatibility [30]	Limited range of materials suitable for nanofinishing	Development of newnanofinishing-compatible materials; surface modification techniques for existing materials	Expanded range of implant material options with improved nanofinished surfaces	Exploration of novel materials such as biodegradable polymers, composite materials, and biomimetic coatings
Scalability and manufacturing cost [30]	Challenges in scaling up nanofinishing processes	Optimization of manufacturing processes for large-scale production; cost-effective nanofinishing techniques	Cost-efficient production of nanofinished implants for widespread use	Development of scalable nanofinishing technologies and equipment; cost-benefit analysis of nanofinishing in comparison to conventional finishing methods

(*Continued*)

TABLE 4.2 (Continued)

Challenge	Limitations	Solutions	Impact on healthcare	Future directions
Long-term stability and durability [30]	Degradation and wear of nanofinished surfaces over time	Development of wear-resistant coatings and protective layers; long-term stability testing protocols	Prolonged implant lifespan and reduced need for revision surgeries	Long-term performance monitoring and clinical studies on the durability of nanofinished implants
Post-implant monitoring and assessment [30]	Limited techniques for non-invasive monitoring of implant performance	Integration of biosensors and wireless communication for real-time monitoring; development of imaging techniques for implant evaluation	Enhanced post-implant surveillance and timely detection of complications	Advancement of implant monitoring technologies for early detection of issues and personalized patient care
Patient-specific implant customization [30]	Challenges in manufacturing patient-specific nanofinished implants	Utilization of advanced imaging and 3D printing technologies for personalized implant fabrication	Improved implant fit, comfort, and patient satisfaction	Integration of nanofinishing with personalized medicine approaches; utilization of 3D bioprinting for complex implant structures

4.6.1 Development of new nanofinishing methods

The field of nanofinishing is continuously evolving, and the development of new techniques plays a vital role in enhancing the quality and efficacy of nanofinished surfaces. Researchers are actively exploring novel approaches to achieve superior surface characteristics, such as enhanced smoothness, reduced roughness, and improved biocompatibility. These advancements involve innovative methods for controlling nanoscale features, such as nanoabrasion, electropolishing, and electrochemical machining. Refining and expanding the range of nanofinishing methods makes it possible to tailor the surface properties of implants to meet specific clinical requirements.

4.6.2 Combination of nanofinishing with other surface modification techniques

To further augment the performance of nanofinished surfaces, combining nanofinishing with other surface modification techniques is an emerging trend. This integration allows for a synergistic approach to address multiple aspects of implant functionality. For instance, combining nanofinishing with coatings, such as bioactive coatings, can promote osseointegration and improve implant stability. Moreover, incorporating nanofinishing techniques with antimicrobial surface modifications can help mitigate the risk of implant-related infections. By exploring these synergistic approaches, researchers can unlock new possibilities for enhancing the biocompatibility and functional properties of nanofinished implants.

4.6.3 Impact of Nanofinishing on Implant Longevity, Wear and Tear, and Other Factors

Understanding the long-term performance and durability of nanofinished implants is crucial for their successful integration into healthcare practices. Researchers are actively investigating the impact of nanofinishing on implant longevity, wear and tear resistance, and other factors that affect their performance over time. This involves comprehensive studies to evaluate the structural integrity of nanofinished surfaces, the ability to withstand mechanical stresses, and resistance to corrosion and degradation. By gaining insights into these aspects, optimizing nanofinishing techniques to enhance implant lifespan and reduce the need for revision surgeries is possible.

4.6.4 Future Directions of Nanofinishing

Several exciting directions hold promise for the future of nanofinishing in biomedical implant applications. These include:

- Development of advanced nanofinishing methods that can achieve precise control over nanoscale features, surface energy, and topographical characteristics.
- Integrating nanofinishing with additive manufacturing techniques, such as 3D printing, to fabricate complex and patient-specific implant structures with improved surface properties.
- Exploration of bioactive nanofinishing approaches that promote enhanced tissue integration and stimulate favourable cellular responses for accelerated healing.
- Advancement in nanofinishing techniques for biodegradable implants, enabling precise surface modifications and controlled degradation profiles.
- Utilization of nanofinishing to create multifunctional implant surfaces, incorporating features such as drug delivery systems, sensing capabilities, and localized therapeutic interventions.
- Application of nanofinishing in combination with regenerative medicine strategies to optimize the integration of implants with host tissues and facilitate tissue regeneration.

By pursuing these future directions, the field of nanofinishing can continue to evolve, leading to significant advancements in the design, manufacturing, and clinical utilization of biomedical implants. These developments can revolutionize patient care, improve treatment outcomes, and drive innovation in the healthcare field.

4.7 CONCLUSION

In summary, this book chapter has delved into the fascinating realm of nanofinishing for biomedical implant applications in healthcare. It has provided insights into the classification of finishing methods and surface characterization techniques, laying the groundwork for understanding this field. By examining the current challenges, including achieving consistent results, clinical translation, biological response, material selection, scalability, long-term stability, post-implant monitoring, and patient-specific customization, we have gained a deeper understanding of the hurdles that must be overcome. Additionally, exploring emerging trends such as the development of new nanofinishing methods, the combination with other surface modification techniques, and the impact on implant longevity and wear has highlighted the potential for transformative advancements in this field. By addressing these challenges and embracing future directions, nanofinishing holds significant promise in revolutionizing healthcare practices and elevating patient care to new heights.

REFERENCES

1. L. Nagdeve, V. K. Jain, and J. Ramkumar, "Nanofinishing of freeform/sculptured surfaces: State-of-the-art," *Manufacturing Review*, vol. 5, p. 6, 2018, doi: 10.1051/mfreview/2018005.
2. M. Montazer and T. Harifi, "Nanofinishing: Fundamental principles," in *Nanofinishing of Textile Materials*, Elsevier, 2018, pp. 19–34.
3. A. Babbar, Y. Tian, V. Kumar, and A. Sharma, "3D Bioprinting in biomedical applications," in *Additive Manufacturing of Polymers for Tissue Engineering*, Boca Raton: CRC Press, 2022, pp. 1–16.
4. V. Kumar, C. Prakash, A. Babbar, S. Choudhary, A. Sharma, and A. S. Uppal, "Additive manufacturing in biomedical engineering," *Additive Manufacturing Processes Used for Biomedical Engineering*, pp. 143–164, 2022, doi: 10.1201/9781003217961-8.
5. A. Thakur, A. Kumar, S. Kaya, R. Marzouki, F. Zhang, and L. Guo, "Recent advancements in surface modification, characterization and functionalization for enhancing the biocompatibility and corrosion resistance of biomedical implants," *Coatings*, vol. 12, no. 10. 2022, doi: 10.3390/coatings12101459.
6. A. Babbar *et al.*, "Additive manufacturing for the development of biological implants, scaffolds, and prosthetics," *Additive Manufacturing Processes Used for Biomedical Engineering*, pp. 27–46, 2022, doi: 10.1201/9781003217961-2.
7] V. Kumar, A. Babbar, A. Sharma, R. Kumar, and A. Tyagi, "Polymer 3D bioprinting for bionics and tissue engineering applications," in *Additive Manufacturing of Polymers for Tissue Engineering*, Boca Raton: CRC Press, 2022, pp. 17–39.
8. S. Chakraborty and V. Kumar, "Development of an intelligent decision model for non-traditional machining processes," *Decision Making: Applications in Management and Engineering*, vol. 4, no. 1, pp. 194–214, 2021, doi: 10.31181/dmame2104194c.
9. V. Kumar, P. P. Das, and S. Chakraborty, "Grey-fuzzy method-based parametric analysis of abrasive water jet machining on GFRP composites," *Sādhanā*, vol. 45, no. 1, p. 106, Dec. 2020, doi: 10.1007/s12046-020-01355-9.
10. H. Bellini, J. Moyano, J. Gil, and A. Puigdollers, "Comparison of the superelasticity of different nickel–titanium orthodontic archwires and the loss of their properties by heat treatment," *Journal of Materials Science: Materials in Medicine*, vol. 27, no. 10, 2016, doi: 10.1007/s10856-016-5767-5.

11. S. Das, B. Sarkar, and V. Kumar, "RIM-based performance evaluation of DLC coating under conflicting RIM-based performance evaluation of DLC coating under conflicting," *Advances in Modelling and Optimization of Manufacturing and Industrial Systems: Select Proceedings of CIMS 2021*, no. February, pp. 303–320, 2023, doi: 10.1007/978–981-19–6107-6.
12. S. Chakraborty, P. P. Das, and V. Kumar, "A grey fuzzy logic approach for cotton fibre selection," *Journal of the Institution of Engineers: Series E*, vol. 98, no. 1, 2017, doi: 10.1007/s40034-017-0099-7.
13. H. El-Hofy, *Fundamentals of Machining Processes: Conventional and Nonconventional Processes, Third Edition (3rd ed.).* Boca Raton: CRC Press, 2018, pp 1–602.
14. H. Lee, H. Kim, and H. Jeong, "Approaches to sustainability in chemical mechanical polishing (CMP): A review," *International Journal of Precision Engineering and Manufacturing - Green Technology*, vol. 9, no. 1. pp. 349–367, 2022, doi: 10.1007/s40684-021-00406-8.
15. H. Lee, D. Lee, and H. Jeong, "Mechanical aspects of the chemical mechanical polishing process: A review," *International Journal of Precision Engineering and Manufacturing*, vol. 17, no. 4. pp. 525–536, 2016, doi: 10.1007/s12541-016-0066-0.
16. A. Sebastian *et al.*, "Electrochemical polishing of two-dimensional materials," *ACS Nano*, vol. 13, no. 1, pp. 78–86, 2019, doi: 10.1021/acsnano.8b08216.
17. S. Zaborski, A. Sudzik, and A. Wołyniec, "Electrochemical polishing of total hip prostheses," *Archives of Civil and Mechanical Engineering*, vol. 11, no. 4, pp. 1053–1062, 2011, doi: 10.1016/s1644–9665(12)60095-8.
18. B. Rauschenbach, "Ion beam figuring and smoothing," in *Low-Energy Ion Irradiation of Materials. Springer Series in Materials Science*, Cham: Springer, vol. 324, 2022, pp. 265–304. https://doi.org/10.1007/978-3-030-97277-6_7
19. V. Pelenovich *et al.*, "Double-step gas cluster ion beam smoothing," *Wuli Xuebao/Acta Physica Sinica*, vol. 70, no. 5, 2021, doi: 10.7498/aps.70.20201454.
20. M. Malaki and H. Ding, "A review of ultrasonic peening treatment," *Materials and Design*, vol. 87, pp. 1072–1086, 2015, doi: 10.1016/j.matdes.2015.08.102.
21. N. Dixit, V. Sharma, and P. Kumar, "Research trends in abrasive flow machining: A systematic review," *Journal of Manufacturing Processes*, vol. 64. pp. 1434–1461, 2021, doi: 10.1016/j.jmapro.2021.03.009.
22. C. Neff, M. Trapuzzano, and N. B. Crane, "Impact of vapor polishing on surface quality and mechanical properties of extruded ABS," *Rapid Prototyping Journal*, vol. 24, no. 2, pp. 501–508, 2018, doi: 10.1108/RPJ-03–2017–0039.
23. A. Abdullah and A. Mohammed, "Scanning electron microscopy (SEM): A review," *Proceedings of 2018 International Conference on Hydraulics and Pneumatics - HERVEX*, Baile Govora, pp. 77–85, 2019.
24. Y. Zhou and J. Du, "Atomic force microscopy (AFM) and its applications to bone-related research," *Progress in Biophysics &Molecular Biology*, vol. 176, pp. 52–66, 2022, doi: 10.1016/j.pbiomolbio.2022.10.002.
25. D. N. G. Krishna and J. Philip, "Review on surface-characterization applications of X-ray photoelectron spectroscopy (XPS): Recent developments and challenges," *Applied Surface Science Advances*, vol. 12. 2022, doi: 10.1016/j.apsadv.2022.100332.
26. T. Zhao and L. Jiang, "Contact angle measurement of natural materials," *Colloids Surfaces B Biointerfaces*, vol. 161, pp. 324–330, 2018, doi: 10.1016/j.colsurfb.2017.10.056.
27. G. Strnad, N. Chirila, C. Petrovan, and O. Russu, "Contact angle measurement on medical implant titanium based biomaterials," *Procedia Technology*, vol. 22, pp. 946–953, 2016, doi: 10.1016/j.protcy.2016.01.094.
28. A. Dutta, "Fourier transform infrared spectroscopy," *Spectroscopic Methods for Nanomaterials Characterization*, vol. 2, pp. 73–93, 2017.

29. A. Barman and M. Das, "Nano-finishing of bio-titanium alloy to generate different surface morphologies by changing magnetorheological polishing fluid compositions," *Precision Engineering*, vol. 51, pp. 145–152, 2018, doi: 10.1016/j.precisioneng.2017.08.003.
30. T. S. Bedi, R. Kant, and H. Gurung, "Advanced finishing processes for biomedical applications," in *Advanced Micro-and Nano-manufacturing Technologies: Applications in Biochemical and Biomedical Engineering*, Singapore: Springer, 2022, pp. 105–126.

5 Understanding the role of nano-micro surface topography on osseointegration and antibacterial properties

Deepak Patil

5.1 INTRODUCTION TO THE ROLE OF SURFACE TOPOGRAPHY IN THE BIOMEDICAL FIELD

The surface of the biomaterial is the first region that comes into contact with the physiological environment; hence, the initial biological response depends on interfacial contact and the improvement of the surface properties by modification of this surface. From this perspective, the field of bioinspired surfaces fits well into the general field of biomimetic materials. The porosity of bone has been tried to be replicated in biomaterials for better osseointegration and till date researchers are working on it [1]. Additionally, extensive investigation of biological surfaces in recent years has revealed that these surfaces display unusual properties; for example, very high water-repelling rough surfaces use the "lotus effect," surfaces with high and adjustable adhesion use the "gecko effect," and non-reflective surfaces use the "moth-eye effect," among others [1, 2]. It was highlighted that special surface wettability stems from a coordinated action between multiscaled surfaces and surface chemistry [3]. The hydrophilicity and hydrophobicity are the first parameters that affect the adsorption of proteins [4]. Hydrophobic surfaces thermodynamically favour the adsorption of proteins from aqueous solutions, but they also induce irreversible adsorption, which can denature proteins. On the other hand, hydrophilic surfaces may prevent proteins from being adsorbed.

Bacterial attachment is a major post-implantation challenge. Bacterial adhesion mainly depends on three main parameters (Figure 5.1); the type of microorganism, the surface material on which the bacterial cell attached, and the surrounding environment [5]. For a manufacturer or surface engineer, the main focus is on the surface properties of implant materials. The nature of biofilm formation depends on surface properties such as surface chemistry, surface topography, and surface wettability [6, 7]. The surface properties of the bacterium also influence its attachment to the

DOI:10.1201/9781003496298-5

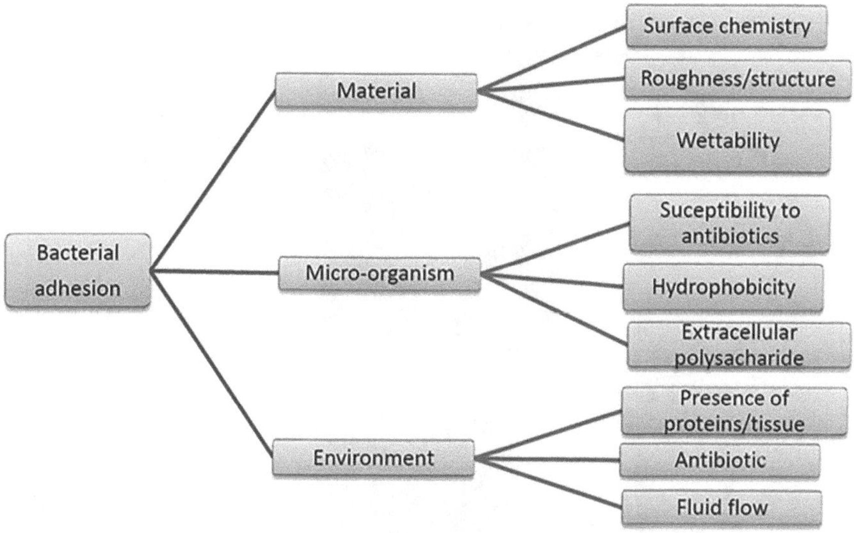

FIGURE 5.1 Flow chart of parameters affecting bacterial adhesion [6].

material surface. It includes the surface charge of the bacterium, the presence of nano-size fimbriae, and flagella on the outer membrane, and the secretion of an extracellular polysaccharide. Among environmental parameters, fluid flow is the dominant factor. Because the microorganism changes its morphology, size, growth rate, and metabolism which resist shear forces are developed due to fluid flow. Therefore, the fluid flow influences the number of attached bacteria and the biofilm structure.

The emergence of new technologies and devices with reduced size requires new materials and surfaces with specific properties, such as low adhesion and friction, and non-wettability, among others. These specific properties will depend on increasing the surface-to-volume ratio so that surface effects will dominate over volume effects [8].

The widely used titanium alloy implants for different applications is depicted in Figure 5.2. Physical and chemical properties, such as size, crystalline structure, charge, composition, and hydrophilicity, determine whether a biomaterial is biocompatible. In other words, whether the best physical and chemical properties are suitable will depend on the application of the biomaterial. For example, cardiovascular articulation implants demand good cell adhesion (Figure 5.2), whereas catheters require that cell adhesion be kept to a minimum to avoid contamination (Figure 5.3). In general, a suitable biomaterial must interact well with body cells and show good cell proliferation; that is, the surface of the biomaterial must be hydrophilic (high wettability), rough (to promote adsorption of proteins), and highly reactive, as well as present high surface energy [4, 9]. In this chapter, we further discuss these aspects of biomaterials, using polymeric, nanoparticulated, and titanium (Ti)-based biomaterials as examples.

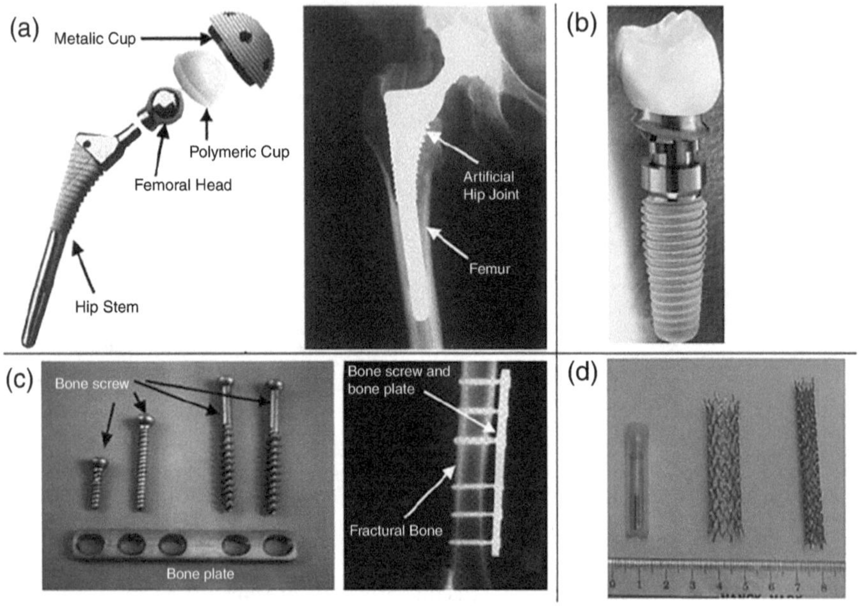

FIGURE 5.2 Biomedical applications of titanium and titanium alloy in the manufacturing of (a) hip joints; (b) dental implants; (c) plates and screw for bone fixation; (d) vascular stents (reproduced with permission from reference [34], copyright @Elsevier).

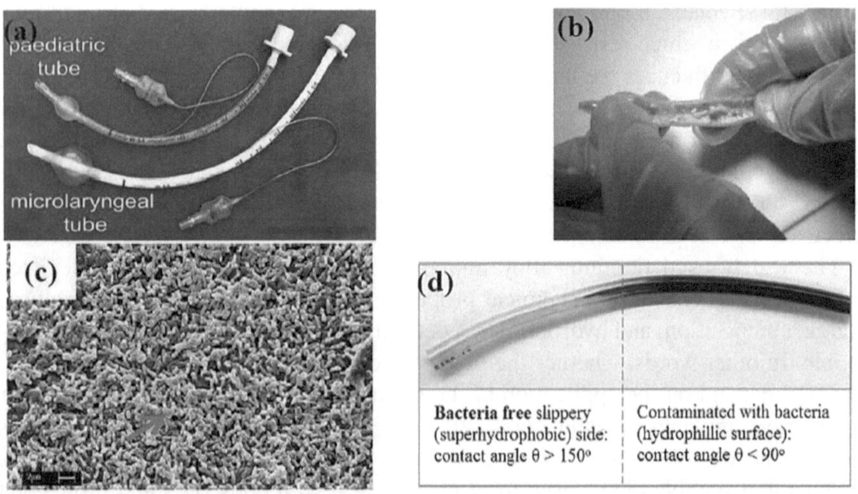

FIGURE 5.3 (a) Biomedical devices, catheters (e.g. paediatric tube, microlaryngeal tube) made of PDMS or polystyrene; (b) the growth of bacteria inside the urinary catheter [12] (c) SEM image of bacterial growth on the catheter surface [12]; (d) effect of superhydrophobic surface on bacterial growth (reproduced with permission from reference [10–12], copyright @ Elsevier).

5.2 EFFECT OF NANOTOPOGRAPHY ON THE GROWTH OF CELLS

Surface wettability plays a vital role in the attachment and growth of cells [13]. The hydrophobic polymer surface of polylactic acid (PLA) is unfavourable for the attachment of mammalian cells; hence, surface treatment to yield terminal -OH groups, for example, renders them hydrophilic and more amenable for cell attachment [13]. However, the excessive -OH group can hinder the growth of mammalian cells. Hence, the most suitable solution is to develop nanostructures on the PLA scaffold to make it hydrophilic and this has been demonstrated in our recent study by using the 3D spin coating technique [14]. Figure 5.4 illustrates the process of generation of nanotopography on a 3D printed PLA scaffold. The nanostructures generated out of the 3D spin coating setup help in increasing surface wettability, which is attributed to the increase in the total surface area caused by nanoscale topographical features on the surface, resulting in a transition from Young's to Wenzel's state of wetting. Unlike surface modification by treatment with NaOH or other harsh reagents, which may be cytotoxic, the nanostructuring presented here utilizes two well-established biocompatible polymers to alter the surface topography at the nanoscale [14].

Figure 5.5(a) shows the optical photos of bacterial colonies formed on PLA-OH and nanostructured-PLA surfaces with the corresponding quantification in CFU/mL in Figure 5.5(b). It was observed that the PLA-OH surface helps with bacterial attachment and growth, whereas the nanostructures create unfavourable conditions for bacterial attachment, which hinders bacterial colonization and growth. The confocal

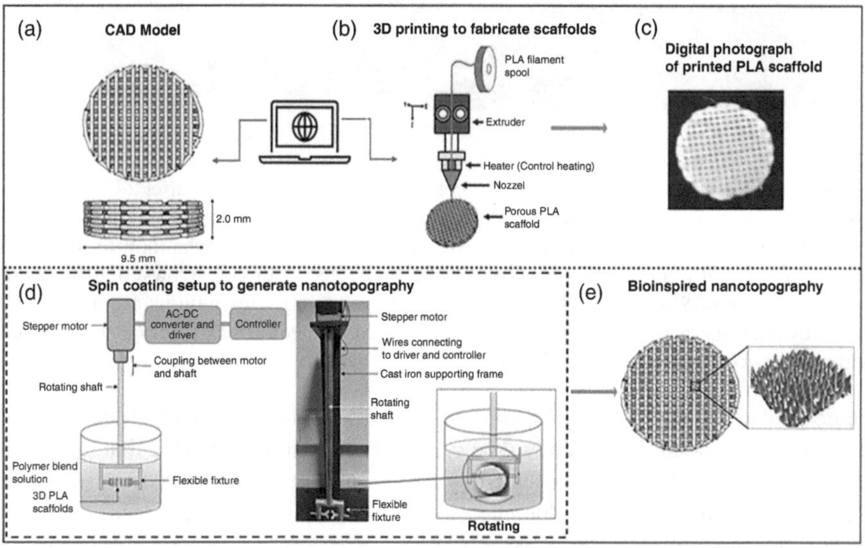

FIGURE 5.4 (a) CAD model for 3D printing with dimensions; (b) schematic representation of FFF printer used for 3D printing PLA scaffolds; (c) photograph of a 3D printed PLA scaffold; (d) schematic illustration of spin coating apparatus for 3D constructs and photograph of the fabricated spin coating apparatus; (e) schematic representing the formation of bioinspired nanotopography on a PLA scaffold after spin coating (reproduced with permission from reference [14], copyright @Elsevier).

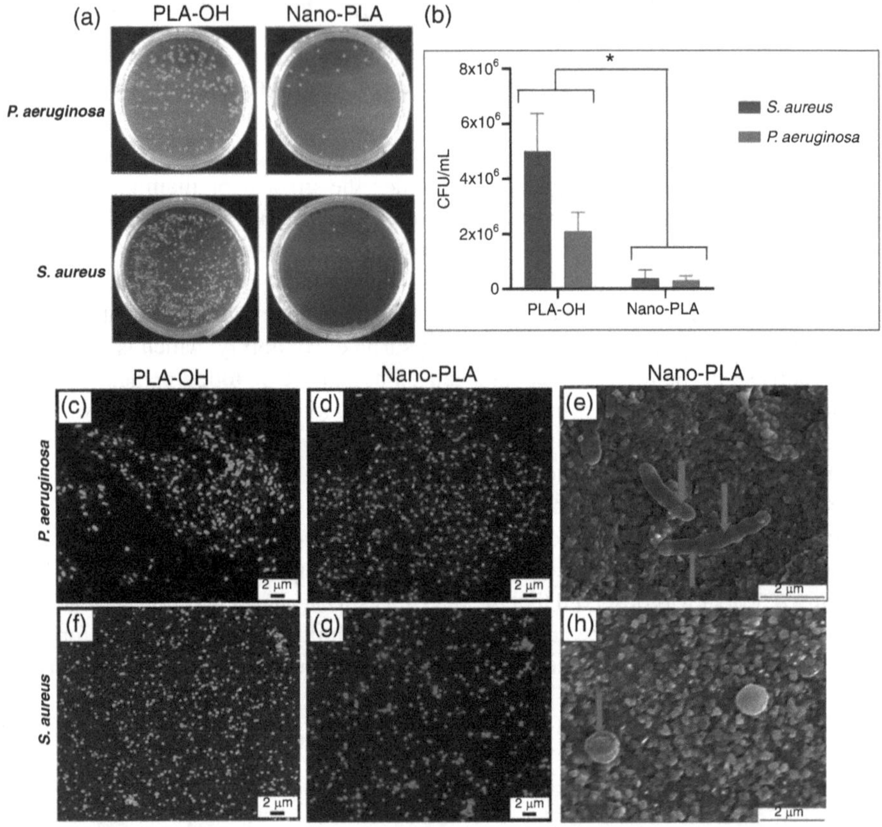

FIGURE 5.5 Testing of *Pseudomonas aeruginosa* and *Staphylococcus aureus* on PLA-OH and nanostructured-PLA scaffolds. (a) The colony formed in petri plates after 24 h of incubation and (b) plot of corresponding CFU/mL. Representative live (green)/dead (red) confocal micrographs of (c, d) *P. aeruginosa* and (f, g) *S. aureus*. SEM micrographs show (e) stretched *P. aeruginosa* with cracks on the outer surface (indicated with red arrows) and (h) damaged membrane of *S. aureus* (indicated with red arrows). * Indicates significance for $p<0.01$ (one-way analysis of variance). The data are presented as mean ± S.D. for $n=3$ (reproduced with permission from reference [14], copyright @ Elsevier).

fluorescent and scanning electron microscopy (SEM) images in Figure 1.5(c) to 1.5(e) further corroborated these trends. The confocal micrographs showed that the nanostructures were bactericidal, revealed by a large number of dead (red) cells, unlike the PLA-OH scaffolds, where the majority of the cells were viable (green).

For potential use as a bone tissue scaffold, it is vital to assess the response of mammalian cells, particularly osteoprogenitor cells, to the nanostructures on the 3D printed PLA scaffolds. Thealkaline phosphatase (ALP) activity was markedly higher on day 21 for both the PLA-OH and the nanostructured-PLA scaffolds as compared to day 7 (Figure 5.6(a)), which suggests that both the scaffolds effectively supported osteogenic differentiation. Notably, cells on nanostructured-PLA scaffolds exhibited markedly higher ALP activity compared to cells on PLA-OH

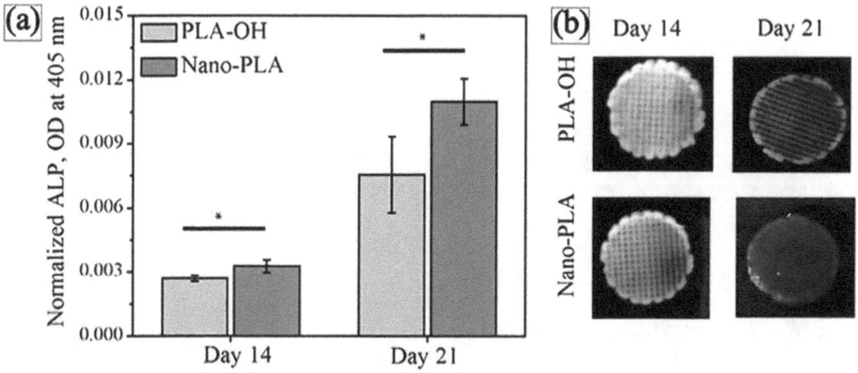

FIGURE 5.6 Osteogenic differentiation of MC3T3-El cells on PLA-OH and nanostructured-PLA 3D scaffolds: (a) quantitative measurement of ALP activity normalized to DNA content on days 14 and 21. Data are plotted as mean ± standard deviation for $n = 3$. For statistical analysis, student's t-test was performed, *indicates a significant difference for $p < 0.05$. (b) Digital images of ARS-stained 3D printed scaffolds on days 14 and 21, wherein the red dye stains for calcium deposits (reproduced with permission from reference [14], copyright @Elsevier).

scaffolds. Taken together, the nanoscale topography on nanostructured-PLA concurrently imparts multi-biofunctional attributes to 3D printed polymer scaffolds, including mechanobactericidal activity, cytocompatibility, and osteogenic activity. In an era of increasing antimicrobial resistance strains, antibacterial yet cytocompatible 3D printed scaffolds may be an effective alternative to biochemical approaches to minimizing infections. Thus, it is envisaged that such a scaffold can promote bone formation while minimizing the risk of implant-associated infections with important implications for the orthopaedic regeneration field.

5.3 NANOTOPOGRAPHY GUIDES THE CELL GROWTH

The focused ion beam (FIB)-SEM images shown in Figure 5.7 shows that the hMSCs cell form abundant filopodia around their periphery upon attachment to the biomedical grade cp-Ti nanopillar (nanostructured-Ti) surface. This is not the case in the cells residing on the flat Ti surfaces (Figure 5.7(a)). The cells had settled on the nanostructured-Ti nanopillars with a top state meaning that the whole cell body was spread on top of the nanopillars. Furthermore, any signs of the deep penetration of the nanopillars into the cell body (i.e. any ruptures or any signs of engulfment by the cell membrane) or local cell stretching in between the adjacent nanopillars were not observed (Figure 5.7(b)).

The high aspect ratio nanostructured-Ti nanopillars have a remarkable potential to direct hMSCs towards osteogenic lineage commitment by manipulating cell adhesion as regulated by focal adhesion kinase (FAK), cell contractility as regulated by Rho-Rho-associated protein kinase (ROCK), and the nuclear translocation of transcriptional factors. Furthermore, these factors are interrelated and can modulate each other's roles. In other words, an osteogenic mechanotransduction pathway can be postulated in which the adhesion and adaptation of the cells to the extreme topography of

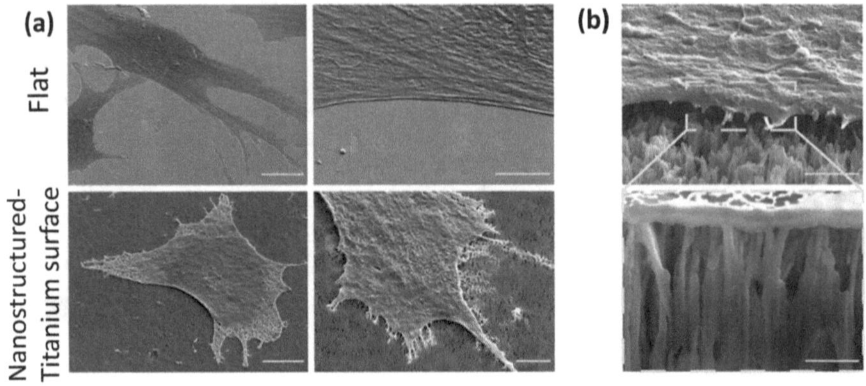

FIGURE 5.7 (a) The representative SEM images of hMSCs settled on the flat Ti and nanostructured-Ti surfaces for one day. Left column: low magnification images showing the cell morphology. Scale bar ¼ 20 μm. Right column: higher magnification images showing the presence (absence) of filopodia around the cell periphery on the nanostructured-Ti (flat Ti) surfaces. Scale bar ¼ 5 μm. (b) A high magnification tilted view of the cell periphery in contact with high aspect ratio nanopillars, scale bar ¼ 1 μm, and a cross-sectional view of the interface (shown by yellow dashed lines), scale bar ¼ 500 nm. The cells have been false-coloured for better visualization (reproduced with permission from reference [15], copyright @ Elsevier).

the surface are highly dependent on FAK, which works in close liaison with ROCK to increase the endogenous forces within the cells and determines the cell shape and area. Whether directly transducing the forces to the nucleus or indirectly enhancing the nuclear translocation of Yes-associated protein (YAP), mechanical stimuli can be translated to biochemical factors, such as RUNX2 in the nucleus (Figure 5.8). Recent advances in the available manufacturing techniques are expected to facilitate the fabrication of 3D implantable structures out of 2D nanopatterned sheets [16, 17]. The availability of such 3D implantable structures would make in vivo studies possible in the future to further investigate similar hypotheses and evaluate the ultimate potential of such meta-biomaterials for bone tissue regeneration.

5.4 SURFACE MODIFICATION TECHNIQUES TO TAILOR THE SURFACES

5.4.1 Surface Texturing

The surface texturing technique has been introduced in implant design to obtain the benefit of the lubricating effect. It was previously used on golf ball surfaces to improve their aerodynamic characteristics. Spiral grooves were produced on bearings to build up pressure to separate the bearing surfaces (Figure 5.9). Surface texturing is now employed in tribology, optics, physics, energetics, biomedicine, electronics, and metrology. This technique is popular in implant design due to the improved friction and tribological performances. It produces micro-textures on implant surfaces that have several benefits over smooth surfaces: (i) they act as a lubricant reservoir [18];

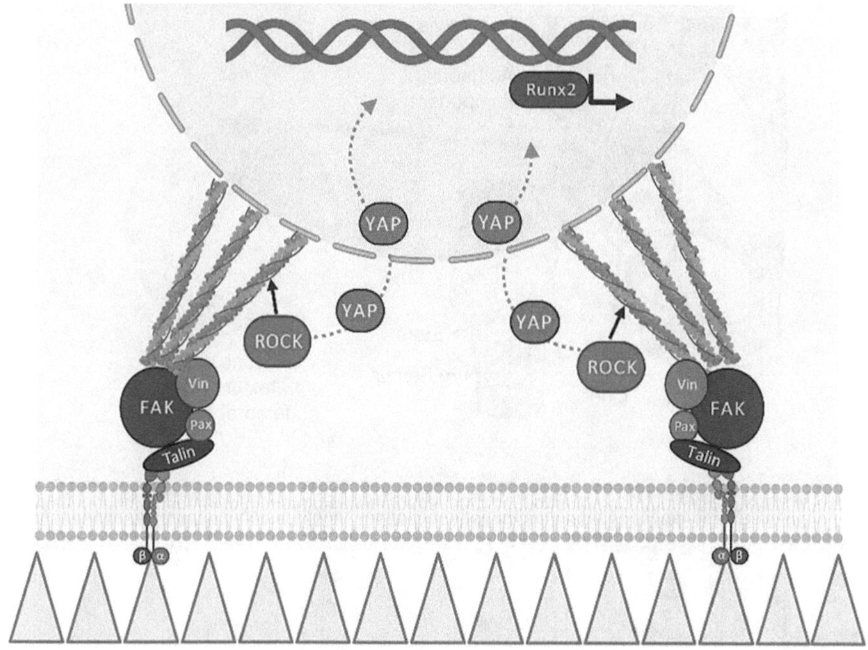

FIGURE 5.8 A schematic illustration of conceivable mechanotransduction pathways triggered by high aspect ratio nanostructured-Ti nanopillars in hMSCs. Cells largely depend on the activity of FAK and ROCK to maintain their normal shape and area on nanostructured-Ti nanopillars. Whether through a direct pathway via focal adhesions (FAs) and cytoskeleton or through an indirect one by enhancing nuclear translocation of YAP, hMSCs may be directed towards osteogenic differentiation indicated by enhanced expression of RUNX2 and mineralization (reproduced with permission from [15], copyright @Elsevier).

(ii) they increase hydrodynamic pressure under sliding conditions [19]; (iii) they store the produced wear debris or foreign materials in dimples [20]; (iv) they decrease the contact area [21]; and (v) they minimize friction and wear [22]. Textured surfaces are more effective, especially in boundary and elastohydrodynamic regimes. Micro-dimple functionality is initially confirmed by correct optimization of its geometrical parameters; still, the influence of dimples quality is not avoidable. Figures 5.9(a) and 5.9(b) show the fabricated surface texturing on articulating femoral heads and SEM images of different types of fabricated textured surfaces.

5.4.2 Surface coating

Improved wear resistance of artificial implants may be accomplished by the application of surface coatings. Surface coating provides morphological (topographical design), physiochemical (changes to surface energy, charge, or composition), and biochemical adaptations (how the cells react) to existing bulk materials. Surface coatings reduce adverse cellular response caused by the generated wear particles. Thus, it engenders the useful life of an artificial implant. Various hard and wear-resistant coatings such

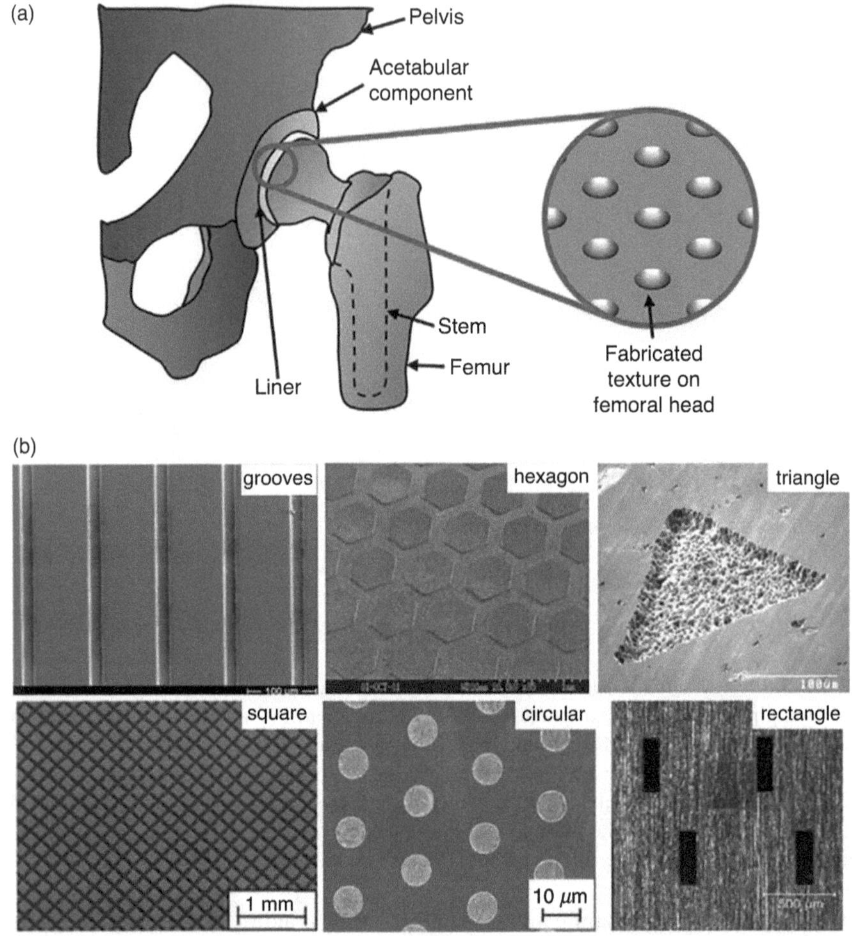

FIGURE 5.9 (a) Surface texturing on femoral head [27] and (b) SEM images of different types of fabricated surface texturing [28] (reproduced with permission from reference [27, 28]).

as metal nitrides, carbides, carbonitrides, and diamond-like carbon (DLC) are used in artificial implant applications. These coatings enhance the mechanical and physical properties such as hardness, elastic strain, and wettability of the surface. The improved wettability of the surface leads to the low friction and consequently the low wear of implant materials. Surface coating protects a bulk surface from tribo-corrosion [23, 24]. The performance of the coated surface also depends on the coating deposition technique. An ideal deposition process can provide quality surface such as a dense homogenous coated surface with excellent adhesion to the substrate.

5.4.3 Surface Grafting

The surface and bulk properties play a significant role in the success or failure of the implant devices. The surface properties such as wettability, chemical composition, softness/stiffness, porosity, and roughness are key factors for the performance of a material in a range of biological environments [25]. Polymers are used as biomaterials. The bulk properties of polymers, such as conductivity, strength, polymer type, stiffness, and general resistance to deterioration, are considered during their selection for biomedical applications [26]. The lower molecular weight proteins start adsorbing on implant surfaces after implantation of biomaterials in a biological environment. A high rate of protein adsorption may result in higher friction. However, protein adsorption depends on chemical composition, surface roughness, and wettability of the surface. Hence, surface modification techniques such as polymer grafting are effective to avoid biofilm formation and protect the surface from biofouling. They also improve wettability of the surface and provide a more hydrophilic surface. The polymer surfaces possess a brush-like hydrophilic structure that is assumed to be similar to that of articular cartilage. The cartilage surface also has a hydrophilic nature due to its water-soluble macromolecules. These hydrophilic macromolecules stimulate the fluid film formation and reduce the friction of the joints [27, 28].

There are two major antibiofouling strategies: The first is to repel the formation of the protein layer and subsequent attachment of bacterial cells and the second is to induce the self-cleaning property in biomedical devices. The formation of the protein layer can be inhibited by modifying the implant surface through a layer of

TABLE 5.1
Summary of surface modification strategies to make antibiofouling surfaces

Methods	Mechanism	Advantages	Disadvantages
	Modification through biological molecules		
Peptide modified surfaces [29]	Structural reformations that inhibit cell adhesion	Super resistance to a wide range of proteins	The high cost of Peptide
	Modification through chemical coatings		
Immobilization of polyethylene glycol (PEG) [30]	Anchoring of PEG using a mussel mimicking linker	Less susceptible to hydrolytic degradation than free PEG	Limited by the availability of suitable surface functionalities for anchoring PEG
Superhydrophilic polymer coating [31]	Zwitterionic polymer coating's resistance to bacterial attachment	Long-term antifouling characteristics	Limited commercial availability of zwitterionic polymers
Superhydrophobic polymer coating [32]	Inhibits the attachment of bacterial cells	Commercially easily available	Toxicity concerns with many hydrophobic polymers
	Modification through surface texturing		
Lotus effect [33,35–38]	Slippery surface	Hierarchical structures reduce wettability	Limited by applicability on a large area
Shark skin patterns [34]	By reducing the drag, bacteria could not attach	Applicable in the marine, aerospace field	Applicable to moving Surfaces

peptide [29]. The common approach adopted for making antibiofouling surfaces is to render the surface superhydrophilic or superhydrophobic. Three broad categories of antibiofouling surfaces are summarized in Table 5.1 with their advantages and disadvantages.

5.5 STATE OF THE ART AND INTERPRETATION

The concept of textured surfaces has emerged from naturally existing textured surfaces. There is a need to study the different kinds of nanotextured surfaces for various applications. The superhydrophobic metallic and polymer-structured surfaces have been found suitable for reducing the attachment of bacterial cells. Laser texturing is one of the best techniques used to make large areas superhydrophobic metallic surfaces. From a fabrication point of view, there is a need to speed up the transformation time from metallic hydrophilic surfaces to superhydrophobic surfaces. Moreover, the bactericidal surface has a comparatively high antibacterial ability. Hence, a new effective method for the development of a bactericidal surface with excellent cytocompatibility needs to be developed. However, extreme care has to be taken to control the dose of bactericidal agents.

The use of chemical treatments to render a surface hydrophobic can lead to its rejection in biomedical devices owing to the violation of safety regulations. Hence, there is a need to establish physical processing routes for the creation of surface structures on polymers. The soft-lithography technique can be explored as an approach for generating structured polymers with superhydrophobic nature. The lithographic techniques (e.g. electron beam lithography, photolithography) are highly suitable for developing a superhydrophobic polymer surface. The direct irradiation of laser on a polymer surface (e.g. polydimethylsiloxane) can alter the chemistry of the surface material and can also facilitate the release of excess silicon. Therefore, the soft-lithography technique has been adopted in this thesis for making structured superhydrophobic surfaces which also gives the flexibility to change the shape and size of structures from the micron to nano level. There is a need to investigate bacterial adhesion from the micron- to nano-level structures.

Considering the size of real-life devices, it is important to have structures on a larger area. The methods adopted for the fabrication of superhydrophobic polymer surfaces are limited to a very small area. Therefore, the low throughput and high machining cost limit the use of these processes for real-life applications. These nanostructures can be replicated on soft materials with the help of a hot embossing technique. There is a need for the development of a hot embossing setup to manufacture micro-nano structures on large areas with minimum processing time.

REFERENCES

1. D. Patil, M. Overland, M. Stoller and K. Chatterjee, Bioinspired nanostructured bactericidal surfaces, *Current Opinion in Chemical Engineering*, 34, pp. 100741, 2021.
2. Z. Li, L. Yin, S. Jiang, L. Chen, S. Sang and H. Zhang, Photocatalytic degradation self-cleaning composite membrane for oil-water separation inspired by light-trapping effect of moth-eye, *Journal of Membrane Science*, 669, p. 121337, 2023.

3. K. S. Liu, A. Fujishima, L. Jiang and M. Cao, Bio-inspired titanium dioxide materials with special Wettability and their applications, *Chemical Reviews*, 114, pp. 10044–10094, 2014.
4. E. A. Vogler, Structure and reactivity of water at biomaterial surfaces, *Advanced Colloidal Interface Science*, 74, pp. 69–117, 1998.
5. D. Viteri, V. S. and E. Fuentes, Titanium and titanium alloys as biomaterials. In: Gegner, J. (ed.) *Tribology — Fundamentals and advancements*. InTech, pp. 155–181, 2013.
6. S. D. Puckett, E. Taylor, T. Raimondo and T. J. Webster, The relationship between the nanostructure of titanium surfaces and bacterial attachment, *Biomaterials* 31, pp. 706–713, 2010.
7. X. Dou, D. Zhang, C. Feng and L. Jiang, Bioinspired hierarchical surface structures with tunable wettability for regulating bacteria adhesion, *ACS Nano*, 9, pp. 10664–10672, 2015.
8. J. Drelich, E. Chibowski, D. D. Meng and K. Terpilowski, Hydrophilic and superhydrophilic surfaces and materials, *Soft Matter*, 7, p. 9804, 2011.
9. H. Chang and Y. Wang (2011) Cell responses to surface and architecture of tissue engineering scaffolds. In: Eberli, D. (ed.) *Regenerative Medicine and Tissue Engineering-Cells and Biomaterials*, 1st edn. InTech, Croatia, pp. 569–588.
10. N. MacCallum, C. Howell, P. Kim, D. Sun, R. Friedlander, J. Ranisau, O. Ahanotu, J. J. Lin, A. Vena, B. Hatton, T. S. Wong and J. Aizenberg, Liquid-infused silicone as a biofouling free medical material, *ACS Biomaterial Science and Engineering*, 1, pp. 43–51, 2015.
11. https://www.howequipmentworks.com/tracheal_tubes/, accessed on 25 June 2023.
12. Complications - Indwelling Catheters, https://www.urotoday.com/urinary-cathetershome/indwelling-catheters/complications/problems.html, published on 14 January 2013.
13. S. Nilawar and K. Chatterjee, Surface decoration of redox-modulating nanoceria on 3D printed tissue scaffolds promotes stem cell osteogenesis and attenuates bacterial colonization, *Biomacromolecules,* 23, pp. 226–239, 2022.
14. D. Patil, S. Kumari and K. Chatterjee, Bioinspired nanotopography on 3D printed tissue scaffold to impart mechanobactericidal and osteogenic activities, *Colloids and Surfaces B: Biointerfaces,* 228, p. 113401, 2023.
15. K. Modaresifar, M. Ganjian, P. J. Díaz-Payno, M. Klimopoulou, M. Koedam, B. C. J. van der Eerden, L. E. Fratila-Apachitei and A. A. Zadpoor, Mechanotransduction in high aspect ratio nanostructured meta-biomaterials: The role of cell adhesion, contractility, and transcriptional factors, *Materials Today Bio*, 16, p. 100448, 2022.
16. T. van Manen, M. Ganjian, K. Modaresifar, L. E. Fratila-Apachitei and A. A. Zadpoor, Automated folding of origami lattices: From nanopatterned sheets to stiff meta-biomaterial, *Small*, 19, p. 2203603, 2021.
17. M. Ganjian, S. Janbaz, T. van Manen, N. Tümer, K. Modaresifar, M. Minneboo, L. E. Fratila Apachitei and A. A. Zadpoor, Controlled metal crumpling as an alternative to folding for the fabrication of nanopatterned meta-biomaterials, *Materials & Design*, 220, p. 110844, 2022.
18. X. Wang and K. Kato, Improving the anti-seizure ability of SiC seal in water with RIE texturing, *Tribology Letters*, 14, pp. 275–280, 2003.
19. I. Etsion, State of the art in laser surface texturing, *Journal of Tribology*, 127, pp. 248–253, 2005.
20. N. P. Suh, M. Mosleh, P. S. Howard, Control of friction, *Wear*, 175, pp. 151–158, 1994.
21. R. Ranjan, D. Lambeth, M. Tromel, et al., Laser texturing for low-flying-height media, *Journal of Applied Physics*, 69, pp. 5745–5747, 1991.
22. S. Ghosh, D. Choudhury, N. S. Das, et al., Tribological role of synovial fluid compositions on artificial joints — a systematic review of the last 10 years, *Lubricant Science*, 26, pp. 387–410, 2014.

23. R. A. Antunes and M. C. L. De Oliveira, Corrosion processes of physical vapor deposition-coated metallic implants, *Critical Reviews™ in Biomedical Engineering*, 37, pp. 425–460, 2009.
24. H. A. Ching, D. Choudhury, M. J. Nine, et al., Effects of surface coating on reducing friction and wear of orthopaedic implants, *Science and Technology of Advanced Materials*, 15, p. 014402, 2014.
25. A. Carré and V. Lacarrière, How substrate properties control cell adhesion. A physical–chemical approach, *Journal of Adhesion Science and Technology*, 24, pp. 815–830, 2010.
26. J. M. Goddard and J. Hotchkiss, Polymer surface modification for the attachment of bioactive compounds, *Progress in Polymer Science*, 32, pp. 698–725, 2007.
27. Y. Dong, P. Svoboda, M. Vrbka, et al., Towards near permanent CoCrMo prosthesis surface by combining micro-texturing and low temperature plasma carburising, *Journal of Mechanical Behaviour of Biomedical Materials*, 55, pp. 215–227, 2015.
28. T. Ibatan, M. Uddin and M. Chowdhury, Recent development on surface texturing in enhancing tribological performance of bearing sliders, *Surface Coating and Technology*, 272, pp. 102–120, 2015.
29. L. Malaeb, P. L. Clech, J. S. Vrouwenvelder, G. M. Ayoub and P. E. Saikaly, Do biological based strategies hold promise to biofouling control in MBRs? *Water Research*, 47, pp. 5447–5463, 2013.
30. B. P. Lee, P. B Messersmith, J. N. Israelachvili and J. H. Waite, Mussel-inspired adhesives and coatings, *Annual Review of Material Research*, 41, pp. 99–132, 2011.
31. S. Kurtz, K. Ong, E. Lau, F. Mowat and M. Halpern, Projections of primary and revision hip and knee arthroplasty in the United States from 2005 to 2030, *Journal of Bone Joint Surgery*, 89, pp. 780–785, 2007.
32. J. Hasan, R. J. Crawford and E. P. Ivanova, Antibacterial surfaces: The quest for a new generation of biomaterials, *Trends in Biotechnology*, 31, pp. 295–304, 2013.
33. A. P. Malshe, S. Bapat, K. P. Rajurkar and H. Haitjema, Bio-inspired textures for functional applications, *CIRP Annals – Manufacturing Technology*, 67, pp. 627–650, 2018,
34. A. Malshe, K. Rajurkar, A. Samanta, H. N. Hansen, S. Bapat and W. Jiang, Bio-inspired functional surfaces for advanced applications, *CIRP Annals – Manufacturing Technology*, 62, pp. 607–628, 2013.
35. D. Patil, S. Aravindan, A. Pal, W. H. Khan, P. Pragya, S. Pundir, I. Xess, S. Mohapatra, V. Perumal and P. V. Rao, Microtopographic superhydrophobic polymer surface to prevent urinary tract infections causing nosocomial drug-resistant bacterial adhesion, *Surfaces and Interfaces*, 41, p. 03239, 2023.
36. D. Patil, S. Aravindan, M. K. Wasson, V. Perumal and P. V. Rao, Fast fabrication of superhydrophobic titanium alloy as antibacterial surface using nanosecond laser texturing, *Transaction of ASME- Journal of Micro- and Nano Manufacturing*, 6, pp. 011002(1–8), 2018.
37. D. Patil, A. Sharma, S. Aravindan and P. V. Rao, Development of hot embossing setup and fabrication of ordered nanostructures on large area of polymer surface for antibiofouling application, *Micro & Nano Letters*, 14, pp. 191–195, 2019.
38. D. Patil, S. Aravindan, M. K. Wasson, V. Perumal and P. V. Rao, Antibacterial and cytocompatibility study of modified Ti6Al4V surfaces through thermal annealing, *Materials Science and Engineering: C*, 99, pp. 1007–1020, 2019.

6 Investigating the material removal mechanism during magnetorheological finishing of extrusion punch surface

Anirudh Chana, Anant Kumar Singh, and Sunil Kumar Paswan

6.1 INTRODUCTION

The extrusion technique is in high demand because it has a significant advantage over rolling for the creation of intricate shapes [1]. In the metal-forming business, this process is expanding quickly. The lessening of friction on the extrusion punches results in lower forces requirement for completion of the extrusion operation [2]. The effects of the controlled friction level on metal flow optimization for operations include forging, extrusion, and rolling [2]. By pressing the billet through a die during the extrusion process, the area of the cross-section of the billet is reduced and given the appropriate shape [3]. Friction restricts the material flow through dies [4]. Temperature variations in the billet and friction between the die-billet container both contribute to non-uniform flow [5]. This friction acting between the surfaces of dies, billets, and surface roughness has a direct impact on extrusion punches.

To reduce the friction of the external surfaces of these components, conventional as well as non-conventional finishing (NCF) processes have been developed. This chapter examines the material removal process used in a magnetorheological (MR) finish that was applied to the exterior of an extrusion punch, a real-world industrial component.

Initially various kinds of conventional finishing (CF) processes such as grinding, lapping, and honing were used for improving the surface finish of the external surface of the extrusion punch. But, the constraints (uncontrolled finishing force, failure to finish upto nano level, introducing various kinds of surface defects) involved in those CF processes necessitated the NCF process for finishing the exterior surface of the extrusion punch. NCF techniques were therefore discovered to

DOI: 10.1201/9781003496298-6

minimize surface roughness up to the nano level [6]. Some of these NCF processes are categorized based on their finishing force control mechanisms [7]. Abrasive flow machining (AFM) [8], chemical mechanical polishing (CMP) [9], elastic emission machining (EEM) [10], and rotational abrasive flow finishing (RAFF) [11] are examples of NCF procedures in which cutting forces cannot be regulated. The forces exerted on the abrasives in these processes are uncontrollable. In contrast, the magnetic field-based finishing method regulates the cutting forces [7]. Magnetic abrasive finishing (MAF) [12], magnetorheological finishing (MRF) [13], MR abrasive flow finishing (MRAFF) [14], ball end magnetorheological finishing (BEMRF) [15], and rotational magnetorheological abrasive flow finishing (R-MRAFF) [16] are a few of these processes that can be used to finish the exterior of cylindrical components. A magnetorheological (MR) solid core rotating tool was utilized by Maan et al. [17] to finish AISI 52100 steel with an HRC of 60–62. In 60 minutes, the surface roughness was decreased from 320 to 20 nm using anMR polishing fluid. Mulik and Pandey [18] finished AISI 52100 steel with a 61 HRC using an MAF setup. With the help of unbounded magnetic abrasive particles, the surface roughness was reduced from 102 to 51 nm in 120 seconds. The MR abrasive honing type finishing procedure was developed by Sadiq and Shunmugam [19] for finishing the external curved surface. The workpieces were put in the holder's cavity and held there by the end cap. In order to complete workpieces made of austenite stainless steel and aluminium alloy, the effects of magnetic flux density on changes in surface roughness were studied. From this research, a magnetic field's threshold limit was discovered.

The current effort focuses on fine finishing the extrusion punch surface that is employed in an industrial tube extrusion process. The H13 die steel material, which has a hardness of 48–50 HRC, is used to make the extrusion punch. The final grinding process in the industry produces a surface roughness of 480 nm. According to the industry standard, the profile utilized for the tube extrusion punch is best for a specific operation. It was produced for the market in accordance with their standards. To further improve its functional uses in the extrusion process, the punch was taken from industry and subjected to the current MR finishing technique with a solid rotating core tool. An experimental setup was created for fine finishing of the exterior surface of the H13 die steel material extrusion punch. The removal of material during the external surface finishing of the extrusion punch has been studied in this work. The experiments were carried out using the experimental setup that Anirudh Chana and Anant Kumar Singh created [20]. The extrusion punch was introduced in the industry after MR finishing to study how a decrease in surface roughness affected its use in the extrusion process.

6.2 MATERIAL REMOVAL MECHANISM

During MR finishing of the exterior surface of the extrusion punch, microchips caused the removal of the substance. This happens by the interfacing of abrasives of the polishing fluid and asperities of cylindrical workpiece's exterior surface. The peaks of the surface roughness are affected by the rotation and reciprocal movement of the abrasive particles as well as the electromagnetic tool core finishing forces. It has been investigated how stresses on abrasive particles affect their ability to remove material. Figure 6.1 shows a schematic representation of the material removing mechanism used in the current investigation to polish the exterior surface of the extrusion punch. This figure illustrates how material is removed from the cylindrical workpiece along the helical path. The combination of three motions—namely, the rotation of the cylindrical workpiece, the rotation of the magnetic tool core, and the reciprocating motion of the magnetic tool core along the axis of the workpiece—leads to the abrasive particles tracking the helical path. Under the influence of the magnetic field, the magnetic iron particles organized themselves into a chain and moved in the direction of the high magnetic field, whereas the abrasive particles moved in the direction of the lower magnetic field. The particles that remove material from the surface of the workpiece are known as active abrasives [21]. The strength of the magnetic flux density in the working gap determines how tightly the iron particle chains that lock the abrasive particles are held together [21]. Greater the gripping of abrasives, better will be the finishing performance.

Figure 6.2(a) represents the condition before collision of abrasive particles and roughness peak on a cylindrical surface. Only the magnetic force acts on iron particles which hold the abrasive. After the abrasive particle and roughness peak collide, Figure 6.2(b) demonstrates that three forces, normal force (Fn), tangential force (Ft), and axial force (Fa), are exerted on the abrasive particles. Due to normal force, iron particles force abrasive particles off of the surface of the cylindrical workpiece. The rotation of the tool and the cylindrical workpiece produces tangential force. The tool's reciprocating motion along the axis of the cylindrical workpiece produces the axial force. The resistance offered by roughness peaks to active abrasive particles against tangential and axial forces. As seen in Figure 6.2(b), the normal force acting

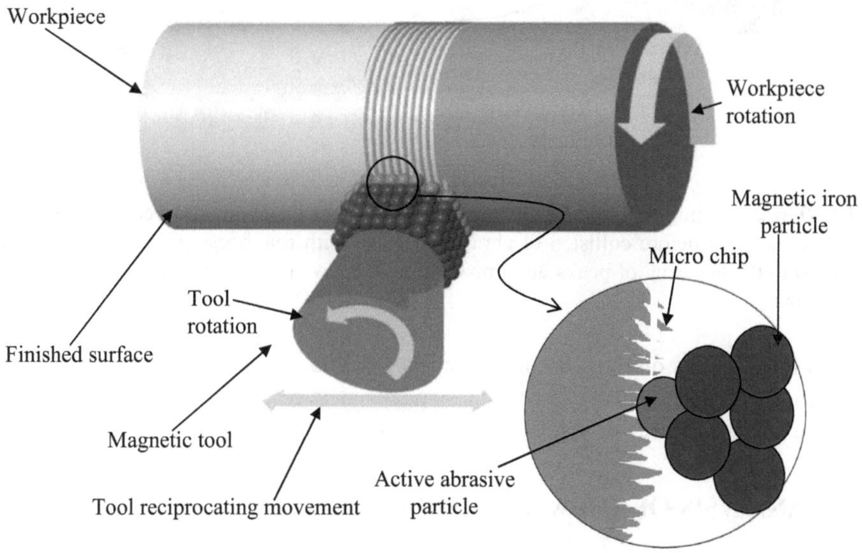

FIGURE 6.1 Mechanism of material removal and formation of helical path on a cylindrical workpiece.

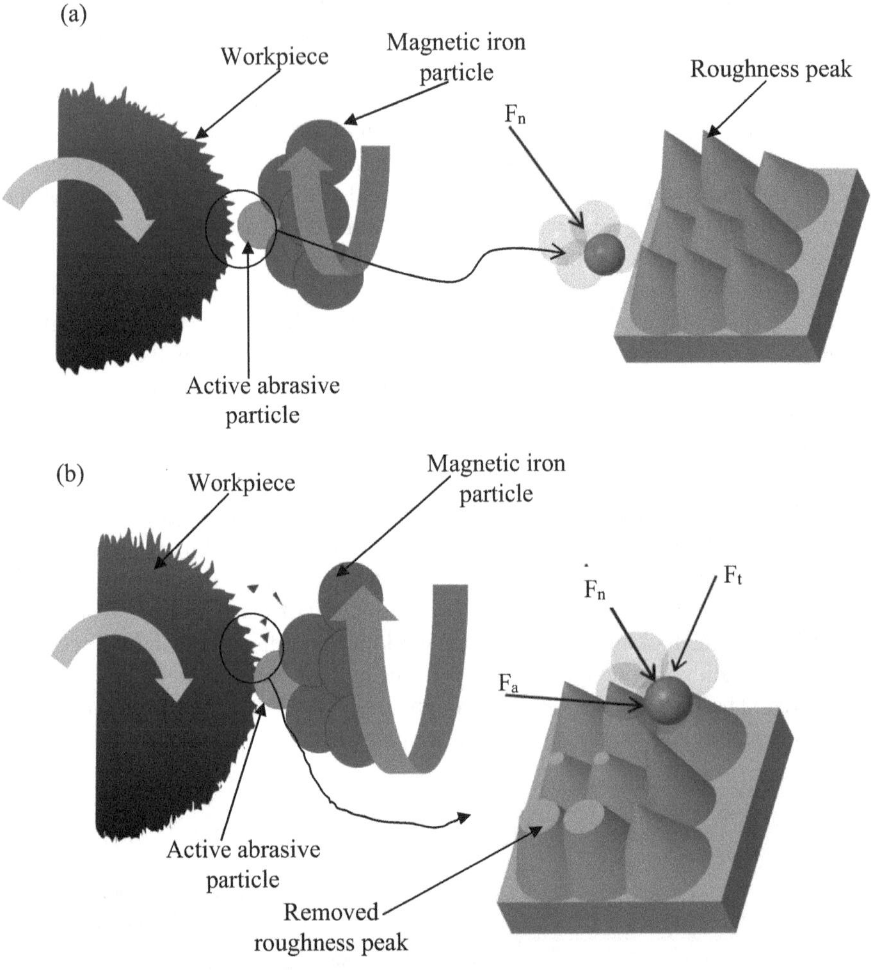

FIGURE 6.2 Acting of forces for the material removal mechanism: (a) force exerting over abrasive particles before collision of abrasive particle with roughness peak and (b) forces involved in the shearing of peaks and impact of an abrasive particle with a roughness peak (side view).

causes indentations on the workpiece's surface, while the product of tangential and axial forces causes shearing of roughness peaks. The resultant must be larger than the resisting shear force of material for removal of peaks to take place.

6.3 ANALYSIS OF THE MR FINISHING TOOL

6.3.1 Magnetostatic analysis of MR finishing tool for the magnetic field

Ansoft Maxwell 13's (student version) magnetostatic simulation was used to verify the viability of the current tool [22]. Magnetic field strength and variation in working gap were checked by performing the simulation. A given model's electromagnetic field problem is resolved by Maxwell Ansoft by designating materials, boundaries, and source circumstances. The Maxwell equations were used in a finite area or space. The model was made in Creo 3.0 and loaded into Maxwell Ansoft to test the magnetic field for the tool's viability. The workpiece and the tool were imported into Maxwell Ansoft. As given in Table 6.1, the attributes were allocated to each model component. The environment was created around the model as shown in Figure 6.3. The model was analysed for ten pass runs. The finite element analysis (FEA) for the CAD model is shown in Figure 6.4. It can be seen that the working gap's magnetic

TABLE 6.1
Input parameters for magnetostatic FEA simulation

Component	Material	Relative permeability
Tool body	Aluminium	1
Coiling	Copper	0.99
Rotating shaft	Mild steel	200
Finishing medium	MR polishing fluid	5
Ferromagnetic cylindrical workpiece	H13 die steel	180

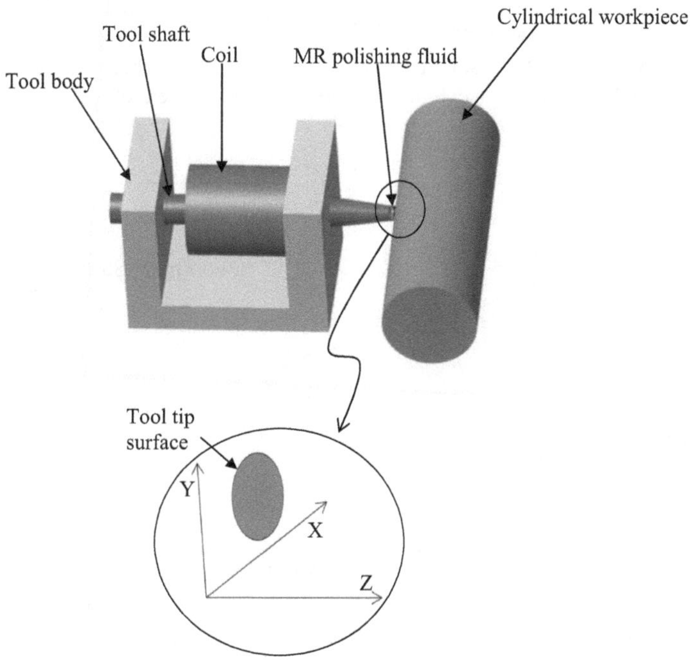

FIGURE 6.3 CAD model prepared for magnetostatic simulation on CREO 3.0 of an electromagnetic tool along with MR polishing fluid, and a cylindrical piece of ferromagnetic material.

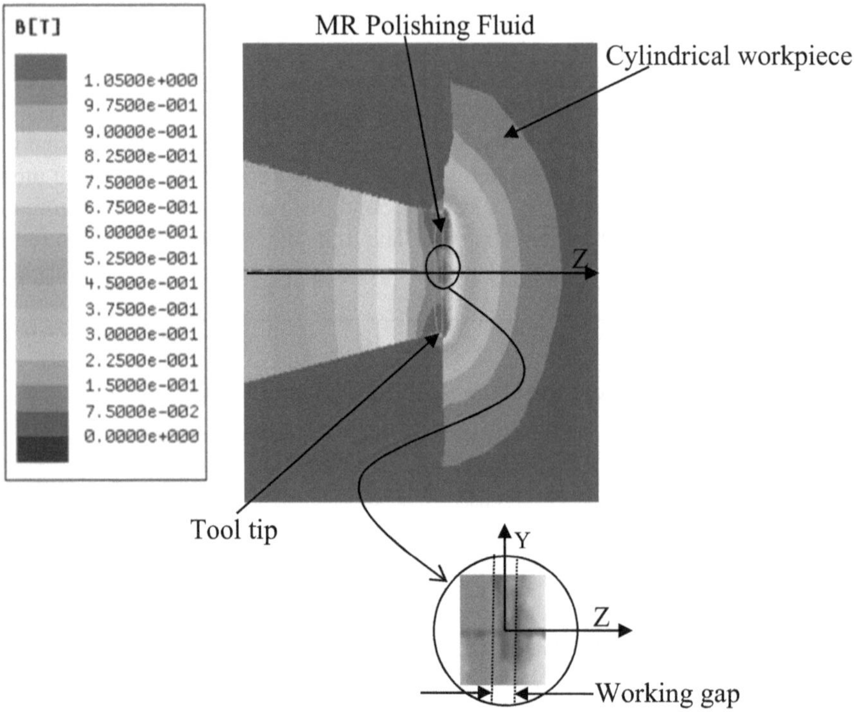

FIGURE 6.4 Magnetostatic FEA image of magnetic field variation between tool tip surface and ferromagnetic cylindrical workpiece and on the tool tip surface.

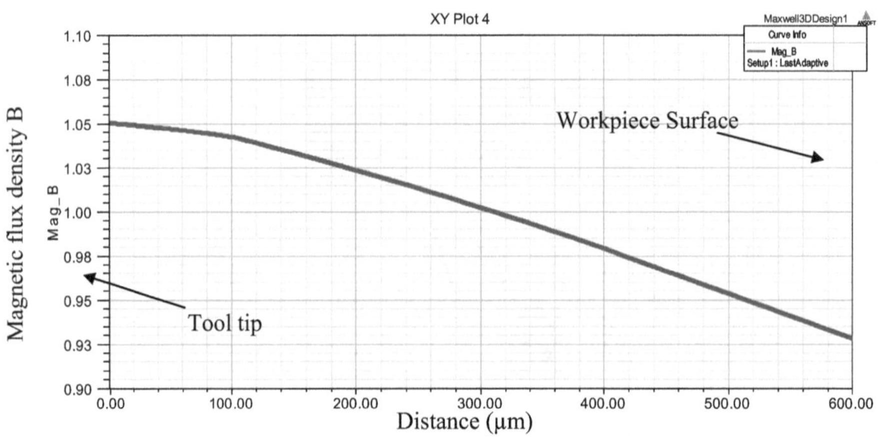

FIGURE 6.5 Magnetic flux density variation in the working gap along the *z*-axis.

flux density ranges from 0.93 to 1.05 T. Figure 6.4 show the graphical representation of variation of magnitude of magnetic field in between the working space. The magnitude of magnetic field on the tool side is 1.05 T and it decreases to the magnitude of 0.93 T towards the workpiece surface, as shown in Figure 6.5. Due to such variation in magnetic field in the functional work space, iron particles move towards a region

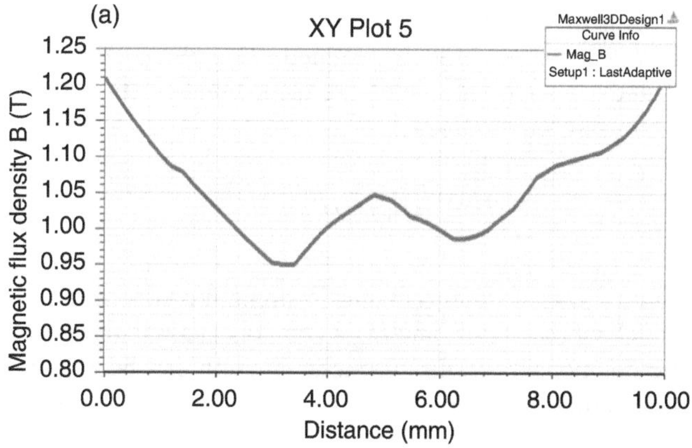

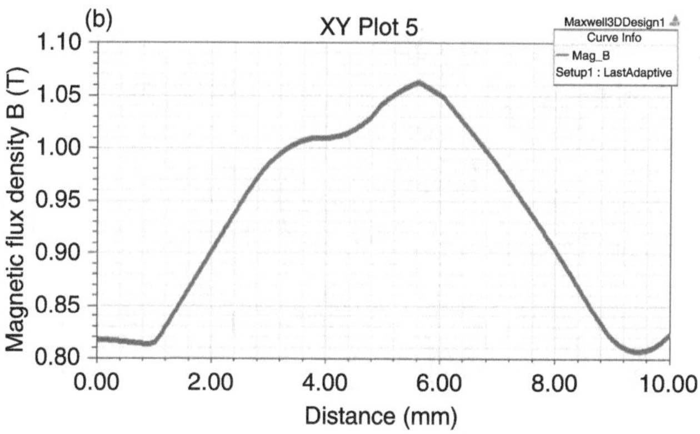

FIGURE 6.6 Magnetic flux density variation on tool face (a) along x-axis and (b) along y-axis.

of high magnetic strength and non-magnetic iron particles move towards a region of low magnetic strength [4]. Figures 6.6(a) and 6.6(b) represent the variation of magnetic flux on the tool surface along the x-axis and y-axis, respectively. It can be seen that the active region for the abrasive is 10 mm along the tool surface. The active abrasive region is in a circular area of radius 10 mm. In this region the magnetic field varies from 0.80 to 1.2 T. This region is accountable for holding the abrasives for finishing the workpiece surface and was used in a later section for the calculation of active abrasive particles.

6.3.2 Analytical Investigation of MR Finishing Tool for the Magnetic Field

Ferromagnetic particles become magnetized by the magnetic field that exists in the working gap. Using Eq. (6.1), the magnetic flux density at any location inside a working gap was computed [23, 24].

$$B = \mu_0 (H + M) \tag{6.1}$$

where H is the magnetic strength at a place, μ_0 is the permeability of free space, B is the magnetic flux density, and M is the magnetization of a particle. The value of H has to be calculated at each point to measure the magnetic field in the gap along the axis of the tool by using Eq. (6.2) [24].

$$H = \sqrt{(H_{xij}^2 + H_{yij}^2 + H_{zij}^2)} \tag{6.2}$$

As seen in Figure 6.7, H_{xij} denotes a magnetic strength component along the x-axis, H_{zij} a component along the y-axis, and H_{yij} a component along the z-axis.

All the three components of the magnetic strength can be calculated by using Eqs. (6.3), (6.4), and (6.5) [24].

$$H_{xij} = \frac{Ia_j^2}{4\left(R^2 + a_j^2\right)^{\frac{3}{2}}}\left(\frac{3xz_i}{\left(R^2 + a_j^2\right)} + \frac{105a_j^2x^3z_i}{8\left(R^2 + a_j^2\right)^3} + \frac{105a_j^2xy^2z_i}{8\left(R^2 + a_j^2\right)^3}\right) \tag{6.3}$$

$$H_{yij} = \frac{Ia_j^2}{4\left(R^2 + a_j^2\right)^{\frac{3}{2}}}\left(\frac{3yz_i}{\left(R^2 + a_j^2\right)} + \frac{105a_j^2y^3z_i}{8\left(R^2 + a_j^2\right)^3} + \frac{105a_j^2yx^2z_i}{8\left(R^2 + a_j^2\right)^3}\right) \tag{6.4}$$

$$H_{zij} = \frac{Ia_j^2}{4\left(R^2 + a_j^2\right)^{\frac{3}{2}}}\left(2 - \frac{3\left(x^2 + y^2\right)}{\left(R^2 + a_j^2\right)} + \frac{15a_j^2\left(x^2 + y^2\right)^2}{2\left(R^2 + a_j^2\right)^2} - \frac{105a_j^2\left(x^2 + y^2\right)^2}{8\left(R^2 + a_j^2\right)^3}\right) \tag{6.5}$$

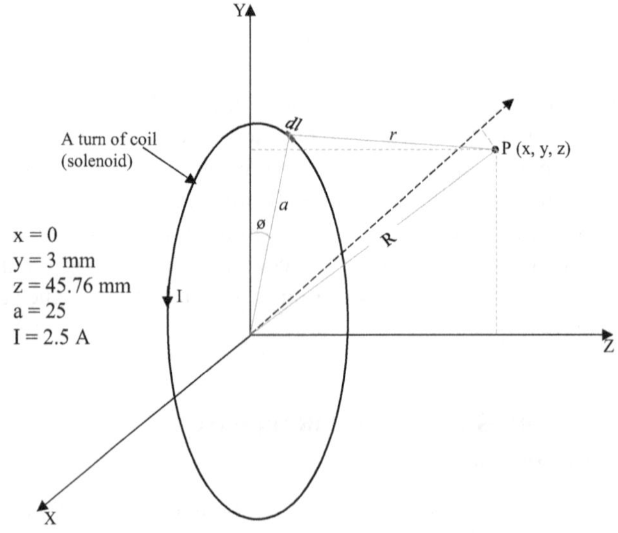

FIGURE 6.7 Co-ordinate system for evaluating magnetic field strength at an arbitrary point.

where I represents the current through the coil of electromagnet (2.5A for present case), a_j is the radial distance of the coil changing in the y direction, z_i is the distance from the measurement point along the z-axis, x and y represent the co-ordinates of measuring points and R, i.e., equivalent to $\sqrt{x^2+y^2+z_i^2}$.

From Eqs. (6.3), (6.4), and (6.5), it can be seen that a_j and z_i vary in the y and z directions, respectively. These two parameters can be calculated by using Eqs. (6.6) and (6.7).

$$a_j = a + jd_g; \quad 0 < j <= n \tag{6.6}$$

$$z_i = z + id_g; \quad 0 < i <= m \tag{6.7}$$

where n is the number of turns radially, m is the number of turns axially, and d_g is the gauge diameter of the coiling wire for an electromagnet, j stands for the number of turns radially in the y direction and i for the number of turns axially in the z direction. In the present tool $n = 26$, $m = 130$, and $d_g = 1$ mm. Magnetic strength of solenoid (H) in the working gap was calculated by using Eq. (6.2) at each magnetic particle in a sole chain and is provided in Table 6.2. The magnetic strength was calculated at the tool face radius of 3 mm along the working gap. A working space of 0.76 mm was calculated for the tool radius of 3 mm between the flat tool and the cylindrical workpiece surface. After obtaining the magnetic strength at each point, the magnetic flux density can be calculated by using Eq. (6.1) in the working gap. The magnetic field can be determined by substituting the value of magnetization in Eq. (6.1) of magnetic iron particles around the working gap. The tool shaft, workpiece, and MR polishing fluid are the ferromagnetic materials in this study, and their respective relative permeabilities are 200, 180, and 5 [25]. The magnetic field is obtained at each point inside the working gap and provided in Table 6.3. As it can be experiential, as given in Table 6.3, magnetic flux density gradient is developed along the working gap. When compared to the surface of the workpiece, the tool surface has a higher density of magnetic flux in terms of magnitude. The changes in magnetic field attained theoretically (Table 6.3) are shown in Figure 6.8(a).

Figure 6.8(b) illustrates how the magnetic flux density alters in the operational gap after a magnetostatic simulation. The magnitude is more in magnetostatic simulation as compared to theoretical because of idle conditions during simulation. Figure 6.8(a) provides the linear equation for the disparity of magnetic flux density in the working gap, which is provided by Eq. (6.8).

$$B(x) = -0.001x + 0.708 \tag{6.8}$$

6.3.3 Analytical investigation of forces involved in finishing action

Furthermore, by comprehending the numerous forces acting on the abrasive particles, the mechanism of material removal in terms of a reduction in surface roughness may be examined. While finishing, a number of forces are at work that lower the roughness peaks. Indentation of abrasives on workpiece surfaces is caused by

TABLE 6.2
Magnetic strength in the working space

Distance from tool surface (mm)	Magnetic strength H (A/m)
0.047	1464.37
0.094	1462.37
0.141	1460.37
0.188	1458.37
0.235	1456.37
0.282	1454.37
0.329	1452.67
0.376	1450.67
0.423	1448.57
0.470	1446.67
0.517	1444.67
0.564	1442.67
0.611	1440.67
0.658	1438.67
0.705	1436.67
0.752	1426.67

TABLE 6.3
Magnetic flux density in the working gap

Distance from tool surface (mm)	Magnetic flux density B (T)
0.047	0.7065
0.094	0.7055
0.141	0.7046
0.188	0.7036
0.235	0.7026
0.282	0.7017
0.329	0.7007
0.376	0.6998
0.423	0.6988
0.470	0.6978
0.517	0.6969
0.564	0.6959
0.611	0.6949
0.658	0.6939
0.705	0.6929
0.752	0.6919

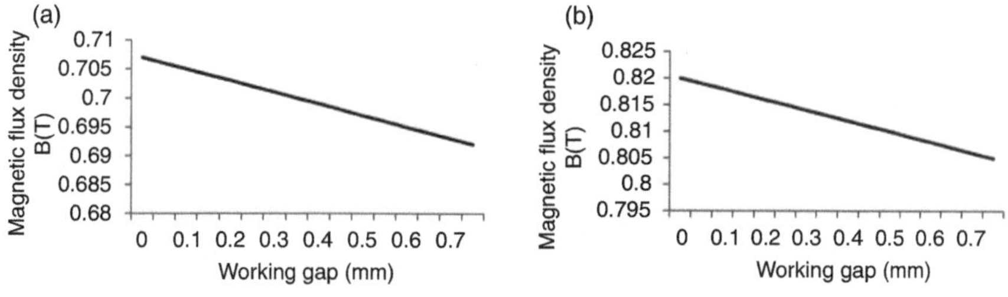

FIGURE 6.8 Magnetic field changes in the working gap (a) calculated theoretically and (b) obtained from magnetostatic FEA.

normal force, while roughness peak shear strength is overcome by tangential and axial forces.

The following assumptions are taken into account:

- Both iron and abrasive particles have a spherical and consistent shape.
- The working gap has a homogeneous distribution of iron and abrasive particles, and the computation of normal magnetic forces does not account for magnetic leakage or losses.
- When using a surface roughness model, calculations are made based on the assumption that the cylinder's surface is uniformly rough.
- The shearing of roughness peaks results from the resultant of tangential and axial force, whereas the indentation of abrasives on the workpiece is caused by the normal magnetic force.
- Because of the low rate of material removal, combining the removed material does not change the properties of the MR polishing fluid.
- In the chain structure of the iron particles, which is continuous in the working gap, are abrasive particles.
- A rounded active zone was considered in order to calculate the helix effect on the active abrasive region.

Figure 6.9 shows the resistant forces when abrasive particles come in contact with the workpiece surface. When the normal force (Fn) and shear force (Fs) acts on abrasive particles, resisting normal force (Rn) and resisting shear force (Rs) occur. The indentation as well as shearing of peaks take place when $Fn > Rn$ and $Fs > Rs$, respectively [23].

6.3.3.1 Unit cell and iron particles in a chain structure

It is crucial to examine how magnetic fields affect both magnetic iron particles and non-magnetic abrasive particles in order to comprehend the forces operating on an abrasive particle. Under the influence of the magnetic field, iron particles develop magnetic dipole moments [4]. It was discovered that magnetic iron particles organize themselves into dipole chains when the forces of dipole interaction outweigh the forces of thermal interaction [4]. Under the influence of the magnetic field, a gradient in the magnetic field is produced in the working gap. Iron particles travel towards high

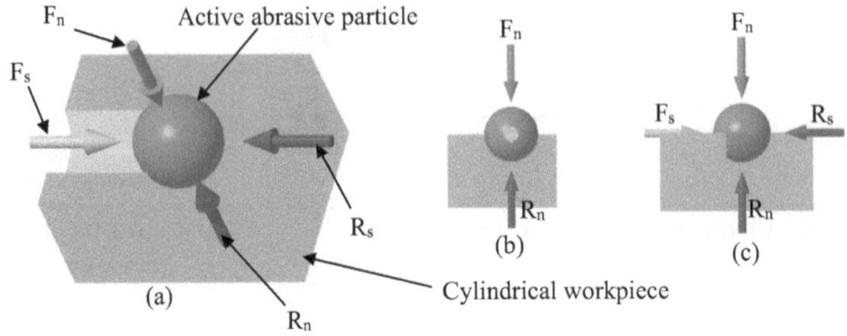

FIGURE 6.9 (a) Different forces acting on a workpiece surface on the interaction of an abrasive particle, (b) side view representing normal forces, (c) front view representing shear as well as normal forces acting on an abrasive particle.

magnetic field gradient regions, while non-magnetic abrasive particles move towards low magnetic field gradient regions [4]. It was discovered that the abrasive particles assemble themselves in the iron particles' continuous chain structure. The form of iron particles was thought to be spherical. Spherical particles are aligned in a magnetic field to produce the chains [26]. The form of abrasive particles is also spherical, and they move very slowly. It was believed that the unit cells are repeated in the MR polishing fluid column in order to calculate the quantity of iron particles. Therefore, the primary goal was to count the number of iron and abrasive particles in a unit cell.

6.3.3.2 Unit cell

In this research, MR polishing fluid consisted of 30% by vol magnetic particles of 400 mesh size and 15% by vol abrasives of 800 mesh size. The diameter of abrasive and iron particles was evaluated by using Eq. (6.9) [27].

$$d_\mu = \frac{15000}{m} \tag{6.9}$$

where d_μ is the diameter of particles in microns and m represents the mesh size of particles. By using Eq. (6.9), it was found that the diameter of iron particles was 38 μm for 400 mesh size and the diameter of abrasives was 19 μm for 800 mesh size.

The number of magnetic particles (N_{IP}) in a certain volume V of MR polishing fluid is given by Eq. (6.10) [26].

$$N_{IP} = \frac{\varnothing_{IP} \times V}{V_{IP}} \tag{6.10}$$

where V is the volume of the MR polishing fluid, $\varnothing_{IP}$ is the volume fraction of iron particles in the MR polishing fluid, and V_{IP} is the volume of a single iron particle. While using the values $\varnothing_{IP} = 30\%$, $V = 1\,mm^3$, and $V_{IP} = 2.873 \times 10^{-5}\,mm^3$ it was found that 10,441 iron particles existed in the MR polishing fluid by using Eq. (6.10). The volume occupied by a single iron particle was $9.57 \times 10^{-5}\,mm^3$ in $1\,mm^3$ MR polishing fluid. Similarly, abrasive particles were 41,767 in mm^3 MR polishing fluid and the volume occupied by a single abrasive was $2.35 \times 10^{-5}\,mm^3$. The edge length of the cubic unit cell was found to be 47 μm. The number of abrasive particles for a single iron article was two. The possible arrangement of active abrasives for a single iron particle is shown in Figure 6.10.

6.3.3.3 Magnetic iron particles in a single chain in the operation gap

The working gap between the tool and cylindrical workpiece is shown in Figure 6.11. The working gap (UQ) was calculated at a radius of 3 mm from the tool face.

PR was calculated from ΔPQR, *as* shown in Figure 6.11, using Eq. (6.11).

$$PR = \sqrt{PQ^2 - QR^2} \tag{6.11}$$

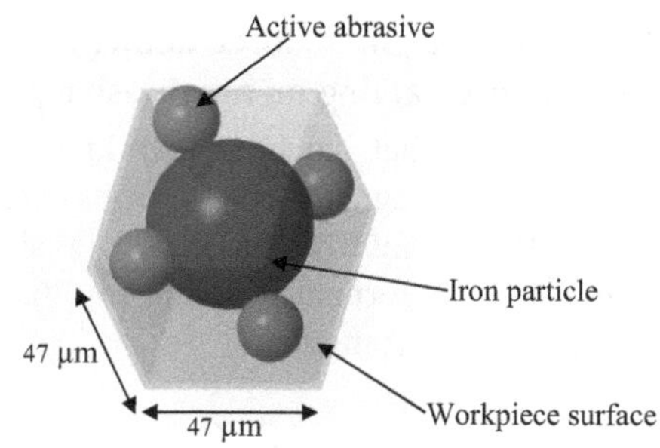

FIGURE 6.10 Possible arrangement of four abrasive particles with the single iron particle in a unit cell.

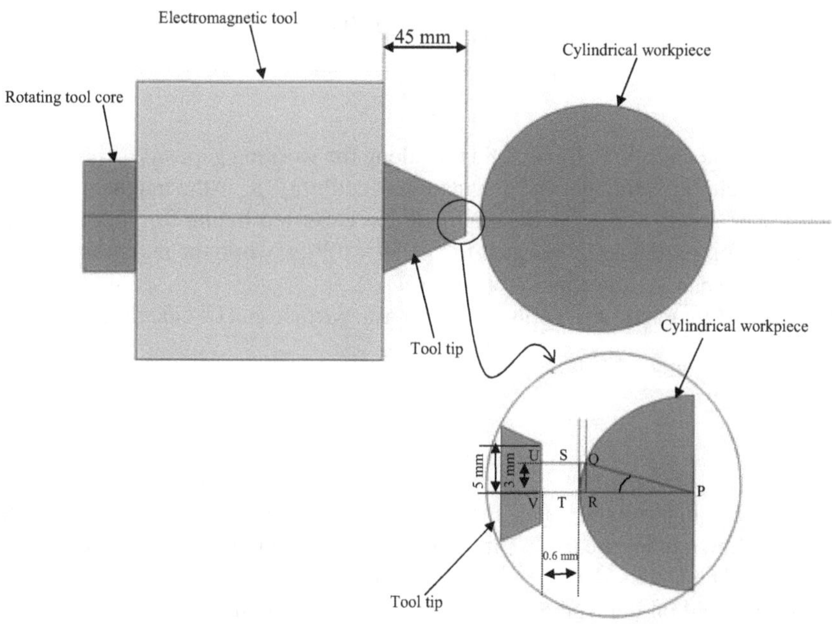

FIGURE 6.11 Schematic for working gap between tool and cylindrical workpiece.

For $PQ = 30$ mm, $QR = 3$ mm, the value of $PR = 29.84$ mm. The working gap (UQ) was calculated from Figure 6.11 using Eq. (6.12)

$$UQ = US + SQ \tag{6.12}$$

$US = 0.6$ mm, $SQ = PT - TR = 0.16$ mm, the value of $UQ = 0.76$ mm. Using Eq. (6.13), it is possible to determine the number of iron particles (N_{IPC}) in the chain at the working gap.

$$N_{IPC} = \frac{\text{Working gap}}{\text{Unit cell edge length}} \tag{6.13}$$

6.3.3.4 Determination of the magnetic force acting on an iron particle and the indenting force acting on an abrasive particle

The active abrasive particles are what indent the workpiece's surface. Under the influence of the magnetic field, the magnetic iron particles exert normal force on the abrasive particles. The force that ordinarily acts on the abrasive particles that cause indentation determines finishing performance. Using Eq. (6.14) [23], the magnetic force exerted on an iron particle was computed.

$$F_m = \frac{m_{IP}\chi_m B\nabla B}{\mu_0} \tag{6.14}$$

Equation (6.14) was further modified to Eq. (6.15).

$$F_m = \frac{m_{IP}\chi_m B(z)\dfrac{dB(z)}{dz}}{\mu_0} \tag{6.15}$$

where $B(z)$ is the variation of magnetic flux along the working gap, m_{IP} is the mass of the iron particle, χ_m is the mass of magnetic susceptibility, μ_0 is the magnetic permeability of free space, and F_m is the magnetic force exerted by the iron particle. The mass of the iron particle, m_{IP}, is equal to 2.3989×10^{11} kg, while the magnetic permeability of free space, μ_0, is equal to 4×10^7 (N/A^2).

The mass of magnetic susceptibility for an iron particle can be calculated by using Eq. (6.16).

$$\chi_m = \frac{\mu_0 M}{B}. \tag{6.16}$$

M was roughly determined using Figure 6.12. The M-B curve for CS grade iron particles is shown in Figure 6.12. Through an experimental vibrating sample magnetometer (VSM) test, the M-B curve was created [12]. Calculating the magnetic force (Eq. 6.15) exerted on the iron particle in the working gap allowed one to determine the indentation force (F_{ind}) acting on the abrasive particle. Table 6.4 lists the magnetic force, that is, F_{ind} can be calculated by using Eq. (6.17) [28]. Eq. (6.17) represents the

TABLE 6.4
Magnetic force acting on iron particles in the working gap

Distance from tool surface (mm)	Magnetic flux density (B)(T)	Mass of magnetic susceptibility $(\chi_m)\times10^{-4}$ (m³/kg)	Magnetic force $(F_n)\times10^{-11}$ (N)
0.047	*0.7065*	3.502	−2.034
0.094	0.7055	3.472	−2.014
0.141	0.7046	3.459	−2.003
0.188	0.7036	3.446	−1.993
0.235	0.7026	3.433	−1.982
0.282	0.7017	3.420	−1.972
0.329	*0.7007*	3.406	−1.962
0.376	0.6998	3.393	−1.950
0.423	0.6988	3.380	−1.941
0.470	0.6978	3.367	−1.931
0.517	0.6969	3.354	−1.921
0.564	0.6959	3.340	−1.910
0.611	0.6949	3.327	−1.900
0.658	0.6939	3.314	−1.890
0.705	0.6929	3.300	−1.880
0.752	**0.6919**	**3.287**	**−1.870**

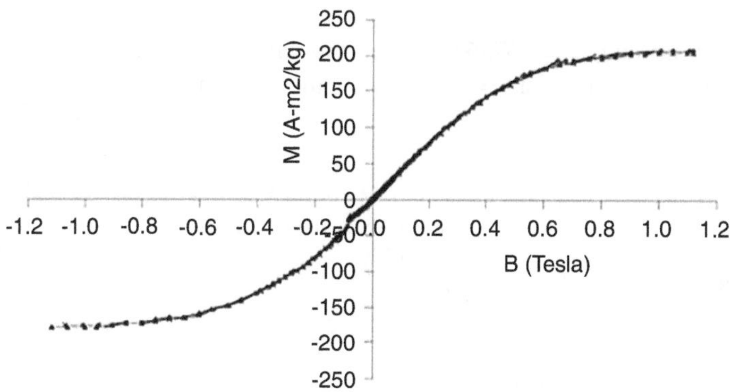

FIGURE 6.12 M-B curve for CS grade iron particles [12].

total force employed by a specific iron particle in the operation gap. The indentation force was equal to 3.15×10^{-10}N.

$$F_{\text{ind}} = \sum_{i=1}^{n} \frac{m_{ip}B_i\left(\frac{dB(z)}{dz}\right)_i \chi_i}{\mu_0} \tag{6.17}$$

where $n = 16$, the number of iron particles in a single chain's working gap exerted on abrasive particles.

6.3.4 Development of Surface Roughness Model

6.3.4.1 Calculating the material removed by a single abrasive

The previous section estimated the indentation force (F_{ind}) acting on abrasive particles. This force is responsible for indenting the abrasives in the workpiece. Indentation causes depression in the workpiece surface, as shown in Figure 6.13(a). The indentation in the workpiece surface is calculated using the Brinell model. The Brinell hardness number (HBHN) is provided by Eq. (6.18) [23].

$$H_{\text{BHN}} = \frac{2F_{\text{ind}}}{\pi D_{\text{abr}}\left(D_{\text{abr}} - \sqrt{D_{\text{abr}}{}^{2} - D_i^2}\right)} \tag{6.18}$$

where F_{ind} is the force used to indent an abrasive particle, D_{abr} is the size of the particle, D_i is the diameter of the indentation, and H_{BHN} is the Brinell hardness number, expressed in kgf/mm^2.

For $F_{\text{ind}} = 3.15 \times 10^{-10}$N, $H_{\text{BHN}} = 400$ Kgf/mm^2 (H13 steel), $D_{\text{abr}} = 19$ μm. The value of D_i from Eq. (6.18) is determined as $D_i = 3.0786 \times 10^{-10}$m. From ΔOGB Figure 6.13(b), the depth of depression can be determined by using Eq. (6.19) [25].

$$t = \frac{D_{\text{abr}}}{2} - \frac{1}{2}\sqrt{D_{\text{abr}}^2 - D_i^2} \tag{6.19}$$

For $D_{\text{abr}} = 19$ μm and $D_i = 3.0786 \times 10^{-10}$m, the value of the depth of depression is calculated using Eq. (6.19) as $t = 4.988 \times 10^{-15}$m. By using the values of D_{abr} and t,

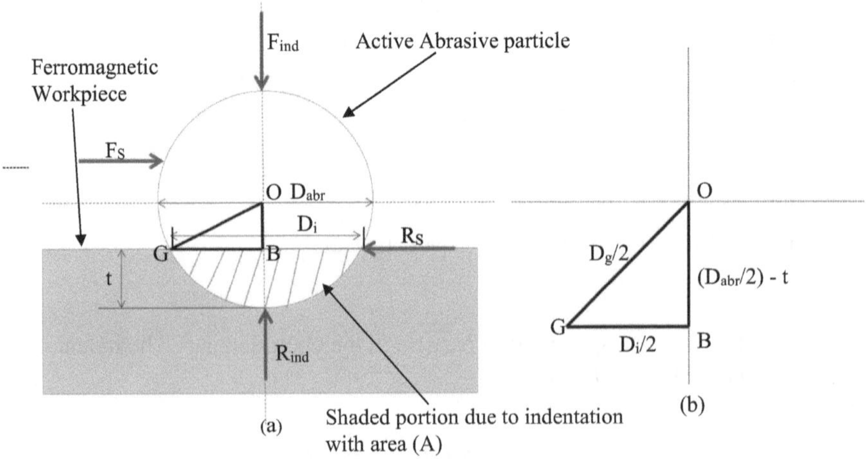

FIGURE 6.13 (a) Forces acting during interaction of an abrasive particle with a workpiece and (b) geometrical diagram for calculation of the depth of indentation.

it was determined from Figure 6.13(a) that the area of the groove created (A) by the abrasive particle on the workpiece surface was $2.92 \times 10^{-15}\,m^2$ [23]. Volume removed by a single abrasive (V_a) is calculated using Eq. (6.20) [25].

$$V_a = A \times L_a \tag{6.20}$$

where A is the area of the cross-sectional indented groove and L_a is the actual contact length by the abrasive particle on workpiece surface.

6.3.4.2 Evaluation of the number of active abrasives

Active abrasives are the abrasive particles that carry out the polishing process. The magnetic flux density at the workpiece surface from the tool tip surface was used to select the active abrasive particles. As mentioned above (Figure 6.4), a magnetostatic FEA was carried out. It was found that a circular area of 10 mm had the minimum magnetic flux of 0.95 T, as shown in Figure 6.4. Magnetic flux density in the area of the active abrasives ranges from 0.95 to 1.2 T. Active abrasives can be calculated by calculating the area of active abrasives ABCD, as shown in Figure 6.14. On the surface of the workpiece, the circular tool tip's projected area creates an ellipse. The area of the ellipse shown in Figure 6.14(b) has to be calculated. From Figure 6.14(b), OP is the radius of the cylindrical workpiece = 30 mm and PC is the projection of the active abrasives region on the workpiece surface = 10 mm. To measure the angle α, Eq. (6.21) was used.

$$\sin \alpha = \frac{PG}{PO} \tag{6.21}$$

For $PG = 5$ mm and $PO = 30$ mm, the value of α was found to be 0.16 rad. The length of arc AC was calculated by using Eq. (6.22).

$$PC = 2\alpha \times PO \tag{6.22}$$

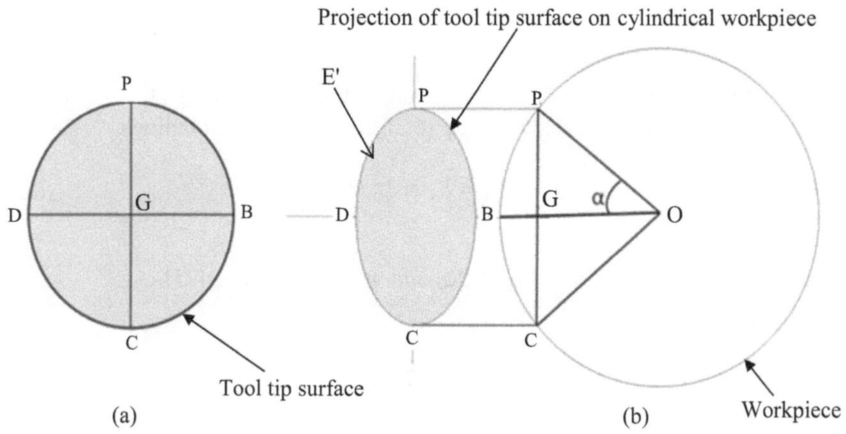

FIGURE 6.14 (a) The area of active abrasives over the tool tip face, and (b) geometrical representation of the region of active abrasives on a cylindrical workpiece surface (side view).

For $\alpha = 0.16$ and $PO = 30$ mm, the length of $PC = 10.04$ mm. Area (E') of the region of the active abrasive (PBCD) is obtained from the expression $\frac{\pi \times PC \times PD}{4}$. For $PC = 10.04$ mm and $PD = 10$ mm, value of $E' = 78.84$ mm^2. For circular tool tip surface area, A' was calculated using Eq. (6.23).

$$A' = \frac{\pi (BD)^2}{4} \tag{6.23}$$

For $BD = 10$ mm, the value of A' calculated using Eq. (6.23) was $A' = 78.53$ mm^2. The difference between A' and E' is not much, so for further calculations A' is used. According to this study, two active abrasives that are positioned at the corners of the unit cell with a length of 47 m interact with the workpiece's surface.

Number of active abrasives (N_s) calculated using Eq. (6.24).

$$N_s = 2 \times \frac{A'}{\text{Area of unit cell face}} \tag{6.24}$$

For $A' = 78.84$ mm^2 and area of unit cell $= 2.209 \times 10^{-3}$ mm^2, Eq. (6.24) was used to determine $N_s = 71{,}100$ as the number of active abrasives.

6.3.4.3 Evaluation of surface roughness reduction

Using Eq. (6.25), the volume removed by active abrasives during the ith stroke (V') was determined.

$$V' = V_a \times N_s \tag{6.25}$$

Actual contact length workpiece width with total height of material removed (V_i) = material removed in ith stroke [25].

$$V_i = L_a \times 2\pi R_w \times h \tag{6.26}$$

where h is the height of the material removed and R_w is the radius of the cylindrical workpiece. Substituting Eq. (6.25) in Eq. (6.26) Eq. (6.27) was obtained.

$$V_a \times N_s = L_a \times 2\pi R_w \times h. \tag{6.27}$$

Replacing the value of V_a from Eq. (6.20) and $h = 7(R_a^{i-1} - R_a^i)$ [25].

$$R_a^{i-1} - R_a^i = \frac{N_s \times A}{7 \times 2\pi R_w} \tag{6.28}$$

where N_s is the number of active abrasives, R_w is the radius of the cylindrical workpiece, A is the indenting area by a single abrasive, R_a^{i-1} is the surface roughness in the (*i*−1)th stroke, and R_a^i is the surface roughness in the *i*th stroke. The material eliminated in a single cycle is represented by Eq. (6.28).

6.3.4.4 Effect of helix angle on region of active abrasives

Two movements are responsible for finishing of workpiece surface, namely, translation motion of tool and workpiece rotation. Due to the combination of these two actions, the material is removed from the workpiece surface in the shape of a helical channel in Figure 6.15(a). The helix angle varies with these motions. The helix angle (β) can be calculated by using Eq. (6.29).

$$\tan\beta = \frac{p}{\pi D} \tag{6.29}$$

where p is the pitch of the helix and D is the diameter of the workpiece. The pitch of the helix (p) can be calculated using Eq. (6.30).

$$p = F \times T' \tag{6.30}$$

where F is the feed of the tool and T is the time taken to complete the path in one rotation along the helix. Time T' can be calculated using Eq. (6.31)

$$T' = \frac{2\pi R_w}{\pi D R} \tag{6.31}$$

where R_w is the workpiece's radius, D its diameter, and R its rotational speed. The pitches for three different combinations of tool feed and workpiece rotation were

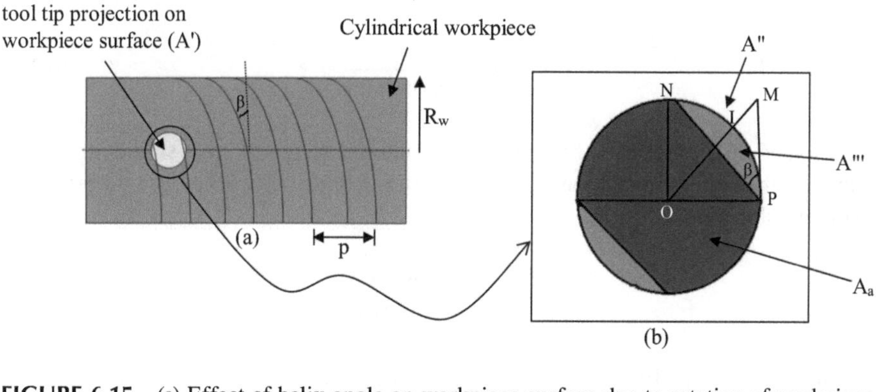

FIGURE 6.15 (a) Effect of helix angle on workpiece surface due to rotation of workpiece and translation motion of tool and (b) schematic of active abrasives region under effect of helix.

TABLE 6.5
Variation of active abrasive region with helix angle

Feed (F) (cm/min)	Workpiece rotation (R) (rpm)	Pitch (mm)	A^a (mm^2)
10	100	1	78.81
30	90	3.33	78.76
20	140	1.428	78.80

evaluated using Eq. (6.30) and are provided in Table 6.5. The area of region PMNIP (A'') can be calculated by using sprandtl Eq. (6.32).

$$A'' = \frac{MN \times MP}{(n+1)} \tag{6.32}$$

where $MP = 5\,\text{mm}$, $MN = MP\tan(\beta) = 5\tan(\beta)$, $n = 2$ for a circle. Figure 6.15 shows variation in the helix angle, which results in the variation of the region of active abrasives. The varying area of active region (A''') can be evaluated by using Eq. (6.33).

$$A''' = \left(0.5 \times MP^2 \times \tan\beta\right) - A'' \tag{6.33}$$

The area of the active abrasive (A^a) varying with helix angle can be calculated using Eq. (6.34) and is given in Table 6.5.

$$A^a = A' - 2A''' \tag{6.34}$$

After calculation of active abrasives region varying with helix angle, the number of active abrasives due to helix effect (N_h) is calculated by using Eq. (6.35).

$$N_h = 2 \times \frac{A^a}{\text{Area of unit cell}} \tag{6.35}$$

Equation (6.36) represents the amount of material removed in a single stroke by adjusting the tool's translation motion and the cylindrical workpiece's rotation, as given in Table 6.6.

$$R_a^{i-1} - R_a^i = \frac{N_h \times A}{7 \times 2\pi R_w} \tag{6.36}$$

As can be seen from Table 6.6, the number of active abrasives does not vary much for different combinations of workpiece rotation as well as tool feed. From Table 6.5, it can be seen that the range of pitch is from 1 to 3.33 mm for the selected range of tool feed and workpiece rotation. The active abrasive region can vary up to a large extent,

TABLE 6.6
Number of active abrasives

A^a (mm²)	Number of active abrasives
78.81	71353
78.76	71308
78.80	71344

TABLE 6.7
Alteration in surface roughness for single stroke

Feed (cm/min)	Workpiece rotation (rpm)	ΔR_a in single stroke (nm)
10	100	0.1575
30	90	0.1578
20	140	0.1578

as shown in Figure 6.15(b); if the helix angle increases, the difference between the number of active abrasives can also increase. Table 6.7 provides the change in surface roughness w.r.t feed and workpiece rotation.

6.4 RESULTS AND DISCUSSION

To determine whether the tool can be used to finish the H13 die steel material, an investigation of the developed tool was conducted. FEA was done on the tool using Maxwell Ansoft (student version). It was found that the magnetic flux of 0.8 to 1.2 T was generated in the working gap. The gradient of magnetic flux density is greatest near the surface of the tool and decreases towards the surface of the workpiece. For the purpose of calculating the magnetic flux density in the working gap, a theoretical model was created. Using Eq. (6.2), the magnetic flux density was computed. The range of the magnetic flux is 0.68 to 0.7 T. The change in magnetic flux gradient in the working gap is comparable to the FEA results. The indenting force was calculated after determining the magnetic flux density. The roughness model was developed for the H13 material. During the MR finishing of a cylindrical workpiece, the theoretical model created for surface roughness was verified experimentally. In this section the results obtained during validation are discussed.

6.4.1 Experimental validation for the surface roughness model

The theoretical model developed was validated experimentally. Experiments were performed on the H13 workpiece. The MR polishing fluid was made in preparation for the experiment. An MR polishing fluid made up of 80% heavy paraffin and 20% grease as the carrier medium, 30% iron particles, and 15% abrasive particles was employed.

TABLE 6.8
Parameters for experimental validation

Experiment	Feed (cm/min)	Workpiece rotation (rpm)
1	*10*	100
2	*20*	140
3	***30***	**90**

TABLE 6.9
Percentage of surface roughness values that are off from their theoretical and experimental values

S. No	Number of cycles	Experimental R_a (μm)	Theoretical R_a (μm)	% Error
0	0	0.380	-	-
1	400	0.320	0.316	1.25
2	800	0.267	0.254	4.86
3	1200	0.206	0.192	6.79

Each set of settings was subjected to a 20-minute experiment. Table 6.8 contains the experiment's parameters. Using the Mitutoyo SJ-400 surf test, the surface finish was measured after the experiment. The theoretical roughness model listed in Table 6.9 underwent tests to ensure its validity. Using Eq. (6.37), the number of cycles was computed.

$$\text{Number of cycles} = \frac{\text{total duration}}{\text{cycle time}} \tag{6.37}$$

The experiment lasted for 20 minutes in total. The tool's motion was calibrated for a 5-mm range. Cycle time was calculated using Eq. (6.38).

$$\text{Cycle time} = \frac{\text{Tool movement}}{\text{Feed of tool}} \tag{6.38}$$

To determine the percentage error between theoretical and experimental data, Eq. (6.39) was employed.

$$\%\text{Error} = \frac{\text{Experimental data} - \text{theoretical data}}{\text{Experimental data}} \tag{6.39}$$

Using the Mitutoyo SJ-400 surf test, the surface roughness was measured after the experiment. Around 0.380 μm was the initial surface roughness measurement (Figure 6.16(a)). It was decreased following testing to 0.320 μm (Figure 6.16(b)) in

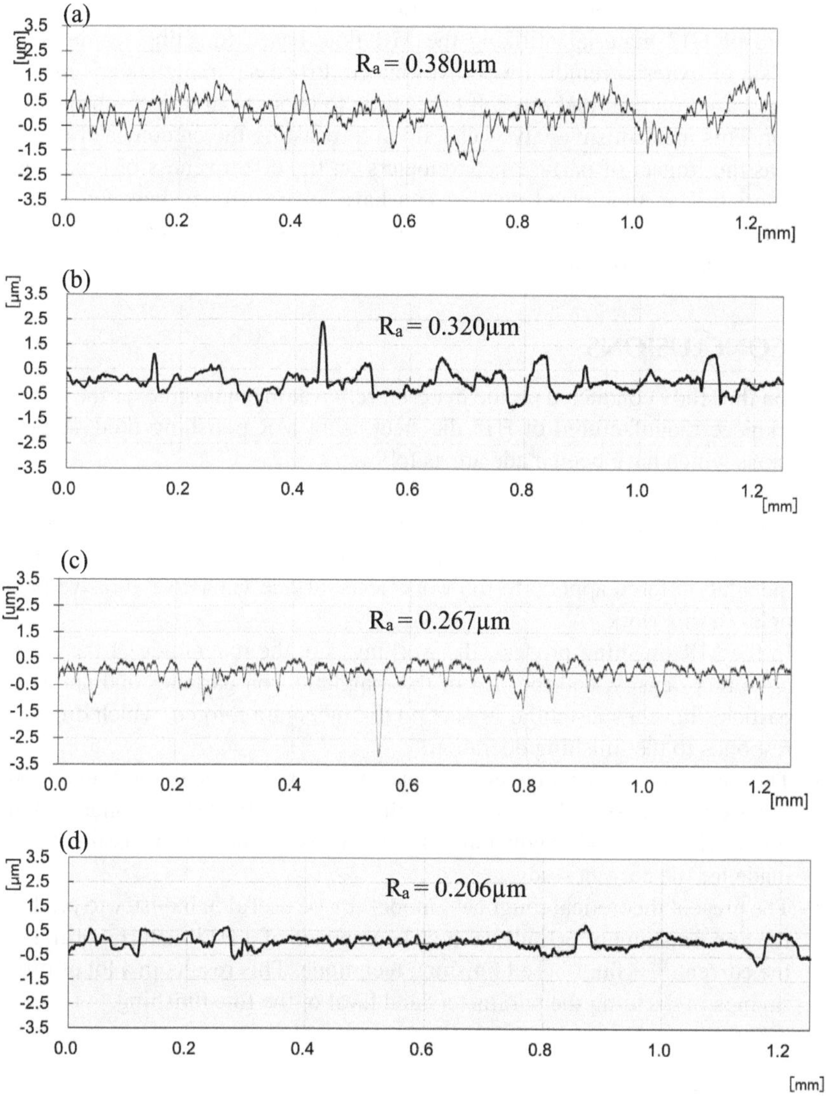

FIGURE 6.16 Surface roughness profile after (a) 0 finishing cycles; (b) 400 finishing cycles; (c) 800 finishing cycles; and (d) 1200 finishing cycles.

400 finishing cycles. On further finishing upto 800 cycles the surface roughness was decreased to 0.267 µm, as shown in Figure 6.16(c)). Lastly, the final surface roughness was achieved as 0.206 µm with 1200 finishing cycles, as shown in Figure 6.16(d). Also, the surface roughness achieved in each set of finishing cycles was as indicated in Table 6.9. The highest error recorded was 6.79%.

To forecast the surface roughness reduction for finishing a cylindrical item composed of H13 material utilizing the MR fluid-based finishing technique, the mechanism of material removal was investigated. Process parameters are chosen in industries based on trial and error. It takes time to choose parameters through trial and error. This mechanism analysis can aid in forecasting the outcomes of finishing as well as the impact of particular parameters on the effectiveness of finishing for H13 material. The anticipated number can have an inaccuracy between 1.25 and 6.79%. By using this mechanism a lot of time can be saved in deciding the parameters for the finishing operation.

6.5 CONCLUSIONS

Based on the study conducted on the material removal mechanism over the cylindrical workpiece manufactured of H13 die steel using MR polishing fluid, some key conclusions which have been made are as follows.

- The trend of variation of the magnetic field between the tool and workpiece linearly decreases towards the workpiece surface, which confirms the indentation force applied to the workpiece surface via active abrasive particles in this work.
- In the MR finishing process, the working gap, the magnitude of the magnetic flux density, and the size of the magnetic iron particles and abrasive particles have a substantial impact on the indentation force, which directly responds to the finishing operation.
- The maximum percentage error between the experimental and theoretical surface roughness (6.79%) advocates that the theoretical study conducted in this work is in good conduct and it is the consequence of the assumptions made for the current study.
- The present theoretical roughness model can be useful in industry to predict the fine finishing of cylindrical items composed of a H13 material utilizing the current MR fluid-based finishing technique. This results in a lot of time savings in deciding the parameters and level of the fine finishing.

REFERENCES

1. Jain, V. K. (2008). Abrasive-based nano-finishing techniques: an overview, *Machining Science and Technology*, 12(3), 257–294.
2. Al-Samarai, R. A., Haftirman, K. R. A., & Al-Douri, Y. (2012). The influence of roughness on the wear and friction coefficient under dry and lubricated sliding, *International Journal of Scientific Engineering and Research*, 3(4), 1–6.
3. Farago, F. T. (1980). *Abrasive Methods Engineering*. New York: Industrial Press Inc.
4. Jain, V. K. (2009). Magnetic field assisted abrasive based micro-/nano-finishing, *Journal of Materials Processing Technology*, 209(20), 6022–6038.
5. Rhoades, L. J. (1988) Abrasive flow machining, *Manufacturing Engineering*, 28, 107–116.

6. Jha, S. and Jain, V. K. (2004). Design and development of the magnetorheological abrasive flow finishing process, *Machine Tool and Manufacture*, 44, 1019–1029.
7. Kordonski, W. I. (1996). Magnetorheological finishing, *International Journal of Modern Physics B*, 10(23–24), 2837–2849.
8. Hayashi, Y., Nakajima, T. and Kunio, T. (2001). Ultra-uniform chemical mechanical polishing (CMP) using a hydro chuck featured by wafer mounting on a quartz glass plate with fully flat water supported surface, *Japanese Journal of Applied Physics*, 35, 1054–1059.
9. Nanz, G. and Camilletti, L. E. (1995). Modelling of chemical-mechanical polishing: A review, *IEEE Transactions on Semiconductor Manufacturing*, 8, 382–389.
10. Komanduri, R., Lucca, D. A. and Tani, Y. (1997). Technological advances in fine abrasive processes, *Annals of CIRP*, 46(2), 545–596.
11. Tsuwa, H., Ikawa, N. and Sugiyama, M. Y (1979). Numerically controlled elastic emission machining, *Annals of CIRP*, 28(1), 193–197.
12. Paswan, S. K., Bedi, T. S. & Singh, A. K. (2017). Modeling and simulation of surface roughness in magnetorheological fluid based honing process, *Wear*, 376, 1207–1221.
13. Pintaude, G. Characteristics of abrasive particles and their implications on wear. Available at https://www.researchgate.net/publication/221912389.
14. Singh, A. K., Jha, S. and Pandey, P. M. (2013). Mechanism of material removal in ball end magnetorheological finishing process, *Wear*, 302, 1180–1191.
15. Singh, M., Singh, A. and Singh, A. K. (2017). A rotating core based magnetorheological nano-finishing process for external cylindrical surfaces, *Material and Manufacturing Processes*, 33, 1160–1168.
16. Maan, S. and Singh, A. K. (2017). Nano surface finishing of hardened AISI 52100 using magnetorheological solid core rotating tool, *International Journal of Advanced Manufacturing Technology*, 95, 513–526.
17. Judal, K. B, Yadava, V. and Pathak, D. (2014). Experimental investigation of vibration assisted cylindrical – magnetic abrasive finishing of aluminum workpiece, *Materials and Manufacturing Processes*, 28, 1196–1202.
18. Jung, B., Jung, K. I., Min, B. K, Lee, S. J. and Seok, J. (2009). Magnetorheological finishing process for hard materials using sintered iron-CNT compound abrasives, *International Journal of Machine Tools and Manufacture*, 49, 407–418.
19. Singh, A. K, Jha, S. and Pandey, P. M. (2013). Mechanism of material removal in ball end magnetorheological finishing process, *Wear*, 302, 1180–1191.
20. Chana, A. and Singh, A. K. (2019). Magnetorheological nano-finishing of tube extrusion punch for improving its functional applications in press machine. *The International Journal of Advanced Manufacturing Technology*, 103, 2037–2052.
21. Jha, S. and Jain, V. K. (2006). Modelling and simulation of surface roughness in magnetorheological abrasive flow finishing (MRAFF) process, *Wear*, 261, 856–866.
22. Maan, S., Singh, G. and Singh, A. K. (2017). Nano-surface-finishing of permanent mold punch using magnetorheological fluid-based finishing processes, *Material and Manufacturing Processes*, 32, 1004–1010.
23. Alam, Z. and Jha, S. (2017). Modeling of surface roughness in ball end magnetorheological finishing (BEMRF) process, *Wear*, 374, 54–62.
24. Xudong, G., Cheng, W. and Ronnguo, Y. (2011). An electromagnetic localization method for medical micro-devices based on adaptive particle swarm optimization with neighbourhood search, *Measurement*, 44, 852–858.
25. Jain, R. K., Jain, V. K. and Dixit, P. M. (1999). Modeling of material removal and surface roughness in abrasive flow machining process, *Machine Tools and Manufacture*, 39, 1903–1923.

26. Yang, A., Yang, H., Yuhang, P., Hongwei, P. and Jinze, P. (2017). Optimum surface roughness prediction for titanium alloy by adopting response surface methodology, *Results in Physics*, 7, 1046–1050.
27. Kirk, D. (2014). A quick and easy formula for mesh-micron particle size conversions. Available at https://www.powderbulk.com/wp-content/uploads/2014/05/pbe 20081201 0018.pdf.
28. Das, M., Jain, V. K. and Ghoshdastidar, P. S. (2018). Analysis of magnetorheological abrasive flow finishing (MRAFF) process, *Machine Tools and Manufacture*, 48, 415–426.

7 Advanced finishing processes for internal cylindrical surfaces
A review

Manish Kumar, Shrushti Maheshwari, and Zafar Alam

7.1 INTRODUCTION

Finishing is considered to be the last stage in the manufacturing process. The surface finish of manufactured components is a critical factor in determining their performance and longevity, as it directly impacts their wear resistance, corrosion resistance, and power loss resulting from friction and strength against crack initiation and propagation. Roughly 15% of the overall manufacturing expenses are allocated towards the completion phase. Evidently, enhancing surface finish even by a unit micron can reflect upon the total cost to a great extent [1].

With the rising requirement for sustainable products with high accuracy, it becomes essential to develop advanced finishing processes. Finishing the internal surface of cylindrical components is becoming a necessity due to increased applications of such components in the aerospace and automobile industry, turbine nozzles, microchannels, bio-implants, cooling channels, surgical tools, and hydraulic manifolds. Owing to the difficulty in nano-finishing internal and complex surfaces of cylindrical components, different available techniques are being modified, and new processes are also being developed.

Traditional finishing procedures (such as grinding, honing, and lapping) are incapable of finishing complex internal free-form surfaces to a high degree of surface finish. To overcome this, various non-conventional processes, viz. abrasive flow finishing (AFF), magnetic abrasive finishing (MAF), and magnetorheological (MR) finishing, are being explored. The present chapter reviews non-conventional processes, their effectiveness in finishing and micro-finishing of internal cylindrical surfaces advances, and hybridization of these processes along with current challenges associated with them.

DOI: 10.1201/9781003496298-7

7.1.1 Need for the finishing of internal cylindrical surfaces

In today's world, accurate super-finish surfaces are in high demand for a variety of industrial components. Pleasant surface topography and reduced surface roughness are characteristics of excellent practical surfaces. Two of the fastest-growing industries that are using components with complicated interior surfaces are the aerospace and automobile industries for gas and fluid flow [2].

Several of the studied abrasive techniques focused on applications that have a sophisticated interior surface with curved characteristics, tiny sections, and variable volumes, such as turbine spray nozzles, hydraulic manifolds, and cooling channels [3, 4].

The interior cylindrical surfaces are finished using conventional methods, including honing and cylindrical grinding. Conventional finishing methods have been observed to present certain apprehensions, such as surface irregularities, excessive heat generation, microcracks formation, residual stress, lack of control over finishing forces, and limited adaptability to various workpiece materials [5]. The majority of die surfaces are characterized by free-form surfaces that require a high degree of precision and polishing. Hence, the typical finishing method is inadequate for the demands of the die surface. Flexibility in finishing different-shaped workpieces is only one of several difficulties that inspired the development of abrasive flow machining (AFM) and rotational abrasive flow finishing (R-AFF).

The utilization of a magnetic abrasive within a magnetic field for the purpose of material removal has led to the emergence of a magnetic field-assisted finishing methodology [6, 7] for creating highly finished surfaces. This method makes use of the adaptability of the abrasive chain arrangement to perform interior finishing of sections where traditional tools are scarcely appropriate, such as inside flexible pipes or complexly bent tubes [3, 8].

Magnetorheological abrasive flow finishing (MRAFF) and rotational magnetorheological abrasive flow finishing (R-MRAFF) are used to smooth the inner cylinder walls of a non-ferromagnetic material through holes. Internal cylinder surfaces of ferromagnetic work material can be finished with a novel magnetorheological honing (MRH) method that employs electromagnets [9]. Electromagnets are employed to create the MRH process. MRH has also been used to improve the finishing performance of cylindrical shaped workpiece of varying diameter [2,5].

7.2 AFF of internal cylindrical surfaces

AFF comprises a slurry consisting of semi-solid abrasive-laden putty mixed into a viscoelastic medium to flow under pressure over a surface to be polished. These abrasive particles are consistently suspended in a flexible polymer carrier. The carrier is a mixture of a rubber-like polymer (polyborosiloxane, known as silly putty) and a lubricating gel [10]. AFF process can be categorized mainly into four categories based on the position and motion of cylinders.

a. One-way AFF: In this sort of framework, the finishing medium is presented at the entrance of the workpiece by a medium chamber and gathered at the exit of the workpiece. The finishing medium is again gathered and reinserted for the next cycle, as shown in Figure 7.1.

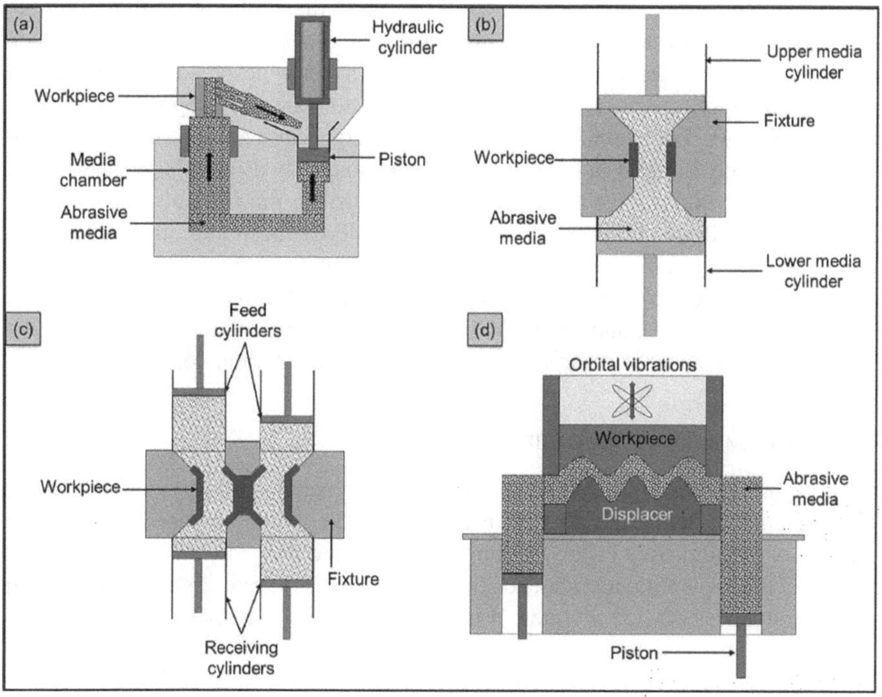

FIGURE 7.1 (a) One-way AFF; (b) two-way AFF; (c) multi-flow AFF; (d) orbital AFF [11].

b. Two-way AFF: In this kind of AFF, the finishing medium ceaselessly responds between two medium chambers and the workpiece is set in the middle of both medium chambers.
c. Multi-flow AFF: This framework is a broadened rendition of a two-way AFF in which another arrangement of two-way AFF is put close to it. The direction of motion of the two arrangements is kept opposite to each other. This setup can work semi- and fully automatically using a computer numerical control (CNC) for mass finishing operations [11].
d. Orbital AFF: Orbital AFF involves moving the medium back and forth between two chambers to achieve final completion. An opposing displacer is used to move the workpiece. A higher-quality surface finish can be achieved by providing the displacer with an extra, pre-set motion that brings the abrasive medium closer to the workpiece.

In orbital AFF the finishing medium moves to and fro between two medium chambers. A displacer is placed opposite to the workpiece. An additional predetermined motion is given to the displacer, which results in better contact between the abrasive medium and the workpiece, leading to a better surface finish.

7.2.1 Significance of AFF in finishing internal cylindrical surfaces

AFF uses a flexible yet controllable medium laden with abrasives as a finishing tool for finishing operations. This leads to its increased capability to reach and finish internal surfaces, and irregular complex contours. Owning to the ability of AFF to precisely finish complex, intricate geometry of the various internal surface, blind holes, deep cavities, thorough holes, etc., it finds its applicability in widespread domains. It has some unique applications, too, where no other customary and progressed finishing process can work. For example, small holes (diameter 3 mm and profundity 30 mm) can be effectively finished in large numbers simultaneously using this process, which is, in any case, extremely challenging. The best surface finishing accomplished by this process is 50 nm [12]. Given such benefits, AFF has been gaining in popularity in expansive applications, especially for accuracy part finishing and complex shape machining; for example, pressure-driven control parts, turbine edges, motor parts, microchannels, nozzles, and so on. AFF can produce a surface finish up to 0.05 μm with close dimensional tolerance of ±5 μm within a short duration of time [13].

7.2.2 Parameters to be Considered in AFF

The prominent process parameters in this technique include viscosity of the medium, the number of cycles of the abrasive medium over the workpiece, the concentration and size of the abrasives in the medium, pressure, and flow rate. As can be seen in Figure 7.2 these parameters influence the material removal rate (MRR) and the final surface roughness of the specimen. It has been observed that coarser grains lead to

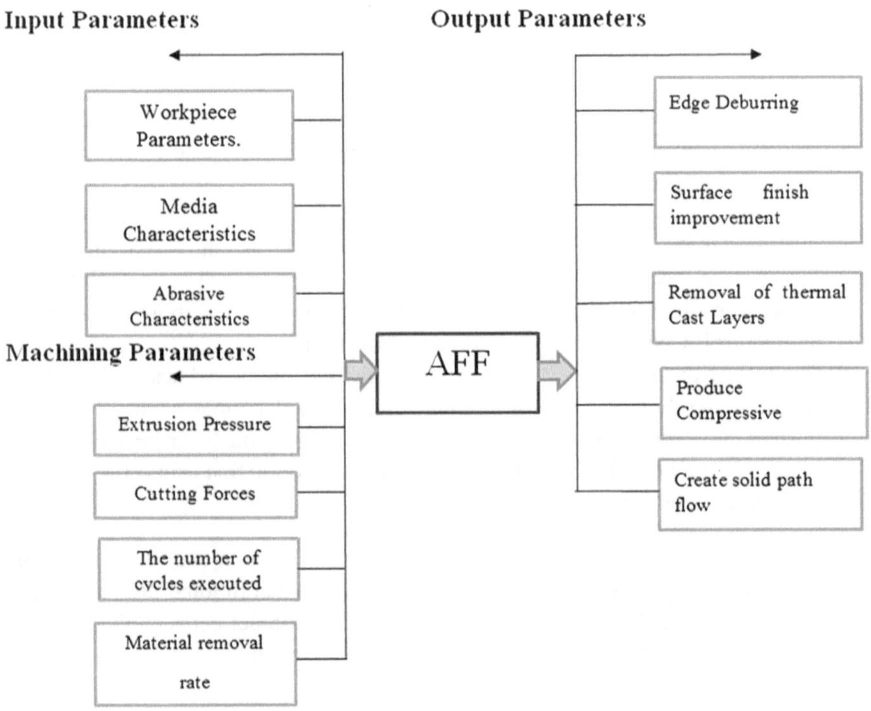

FIGURE 7.2 Parameters in the AFF process.

high MRR but at the expense of surface finish, whereas more refined grains result in better surface finish and low MRR [14]. Similarly, higher viscosity of the medium, increase in pressure, and increase in the number of cycles contribute towards higher MRR. Higher concentrations of abrasives consequently increase the number of active grains in the flowing medium, leading to a better surface finish and MRR [15].

7.2.3 Variants of AFF for Finishing of Internal Cylindrical Surfaces

7.2.3.1 Centrifugal abrasive flow finishing (CFAFF)

The setup includes inserting a spinning rod into the hollow cylinder as the workpiece rotates around it. As the media travels through the workpiece channel and the centrifugal force generator (CFG rod) spins, the abrasive particles are subject to a centrifugal force in the direction perpendicular to the axial force piston. As can be seen in Figure 7.3, the combined impact of these two forces improves the abrasive particles' contact with the workpiece's passage surface. Fewer process cycles are required to complete the workpiece in CFAAFM [16].

The effect of various process parameters has been studied, viz. extrusion pressure, the number of cycles, abrasives grain size, shape, and the rotational speed of CFG rod. It was concluded that (a) the CFG rod significantly increased the effectiveness of AFF by reducing machining time by 70%–80%. (b) Rectangular-shaped rods produced the greatest improvement in surface finish, followed by square, spline, and triangular, whereas spline rods produced maximum material removal. Considering both surface finish and material removal, rectangular CFG gives the best result [17].

The CFG rod's remarkable speed and high extrusion pressure are favourable conditions for increasing surface finish. Then again, a high velocity of the CFG rod, along with a bigger size of rough particles, contributes towards an upgraded MRR [18].

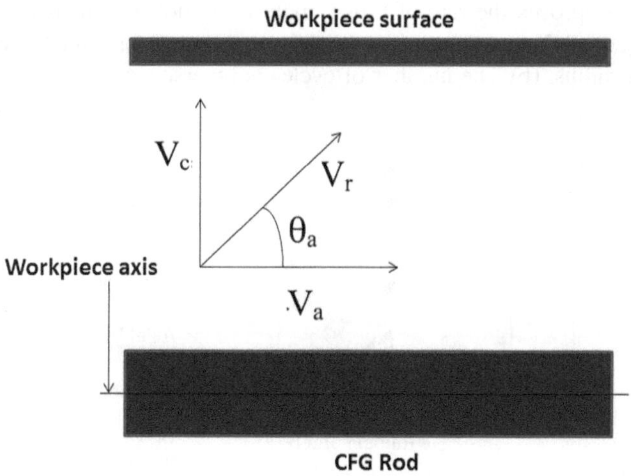

FIGURE 7.3 Schematic of forces in AFF with additional centrifugal force [16].

7.2.3.2 Rotational AFF (R-AFF)

Rotational AFM has been developed and rotational AFF is similar to centrifugal abrasive flow finishing, with the only difference being in the motion of the workpiece. In this technique, the workpiece is situated in the explicitly planned workpiece holder and cinched between the chambers. The two fundamental elements of the workpiece holder are to cinch the workpiece and guarantee a controlled media stream in a shut framework. Inside the installation, the medium courses through a restricting channel prior to arriving at the workpiece's irregularities. At first, the crushing medium is warmed to working temperature by the radiator/cooler. Then, at that point, the crushing medium is squeezed upwards into the workpiece holder along with the machined workpiece shapes. From that point onward, the cycle is rehashed the other way. Rotary motion is provided in the R-AFF workpiece, and the medium oscillates. The combined effect of these two motions works together to improve surface quality while accelerating material removal [19]. As can be seen in Figure 7.4, the abrasive particles in the finishing medium are subjected to not only axial force (F_a) and radial force (F_r), but also tangential force (F_t) due to the workpiece's rotation.

The combined effect of these two motions enhances surface finish while also increasing material removal [19]. Due to the rotational motion of the workpiece, the tangential force (F_t) also acts along with the axial force (F_a) and the radial force (F_r) over abrasive particles contained in the finishing medium, as shown in Figure 7.4.

The surface finish increased by 44% and 81.8% concerning the AFM process. Since the rotation process had a larger drill diameter, it increased the force applied by the medium's flow and hence had greater R_a and MRR [20].

7.2.3.3 Drill bit-guided AFF (DBG-AFF)

Drill bit-guided AFF has been developed and the drill bit changes the flow path of abrasive mediums, increasing the length of contact of abrasives with the workpiece. This, in turn, improves the rate of finish, material removal, and surface texture. It has been suggested that (a) the surface finish is directly proportional to the MRR and the drill bit's radius. (b) The number of cycles is inversely proportional to the area

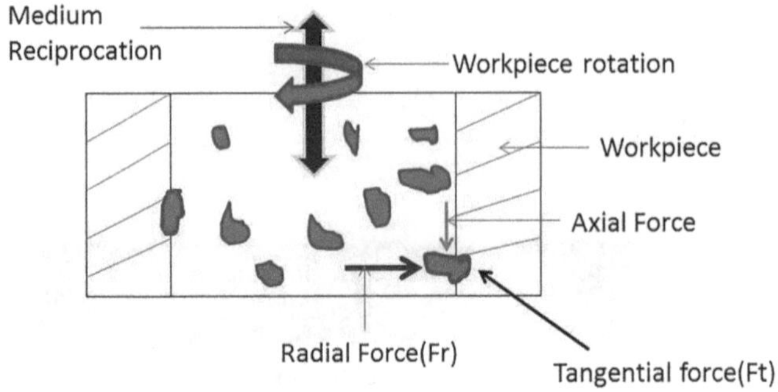

FIGURE 7.4 Schematic of forces during R-AFF [19].

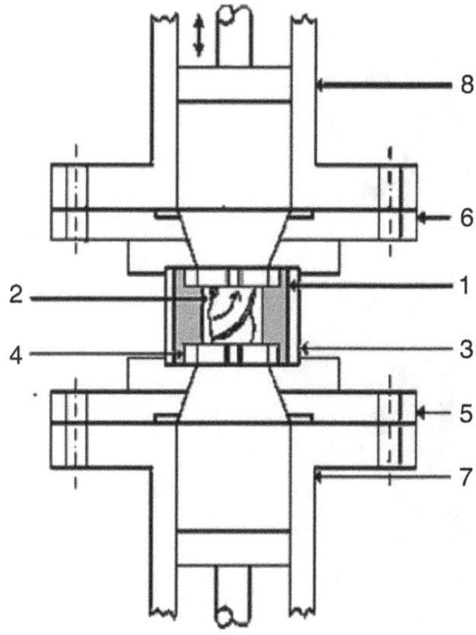

FIGURE 7.5 Schematic representation of drill bit-guided AFF (DBG-AFF) [22].

of abrasives and the roughness of the surface. The drill bit helps in the reduction of peaks and thus improves the quality of finishing. In the drill bit the path travelled by the abrasive medium is longer than that in AFM [21].

The DBG-AFF technique utilizes a twin slot fixture plate to aid in the alignment of the drill bit within the finishing zone along with the assembly of all tool parts, namely: (1) workpiece, (2) drill bit, (3) fixture barrel, (4) twin slotted fixture plate, (5) lower taper fixture, (6) upper taper fixture, (7) lower medium cylinder, (8) upper medium cylinder. Upon entering the finishing zone, the cylindrical slug undergoes a division into two halves. Subsequently, upon exiting the zone, these two halves reunite and undergo a process of remixing. The extent of abrasive intermixing in DBG-AFF is influenced by two factors: the deformability of the medium itself, similar to AFF, and the pressure exerted by the drill bit. Additionally, there are three distinct types of flows that take place in the finishing zone as well as the remixing of the medium upon exiting this zone. The presence of various flows results in a curved workpiece-abrasive contact length, as opposed to a linear one. Consequently, the increased curvature leads to an augmented number of peaks that are susceptible to shearing as shown in Figure 7.5 [22].

7.2.3.4 Helical abrasive flow finishing (H-AFF)

The AFF process was modified using a drill bit. This modified finishing process was termed as the helical abrasive flow finishing process (H-AFF). A helical drill bit was placed co-pivotally with the workpiece in this process [23]. The medium was forced to course through the annular space between the fixed helical drill bit and the workpiece so that more grating grains interface with the work surface. The workpiece used was a hollow cylinder made up of brass, and the abrasion process finished the internal surface of this cylindrical workpiece. AFM leaves

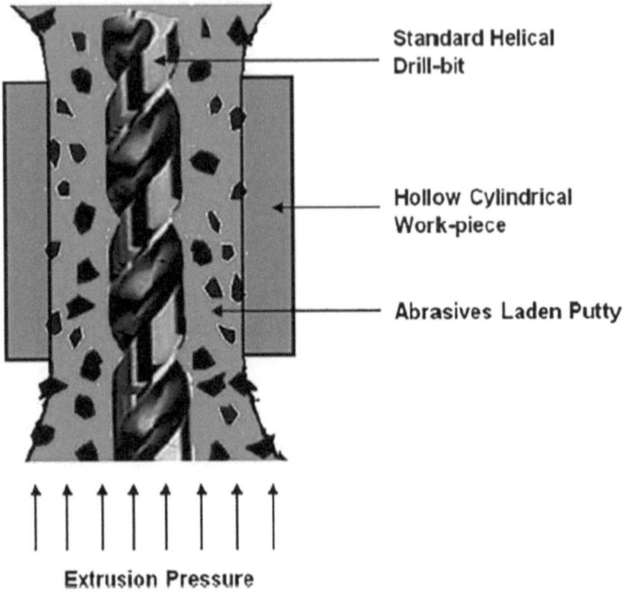

FIGURE 7.6 H-AFF process [23].

abrasives marks perpendicular to the surface; thus, complete removal of irregularities does not occur. As shown in Figure 7.6, on the contrary, H-AFF abrasives make further scratches and follow a helical completing way along with the flute of the helical drilling tool, bringing about the total expulsion of the slopes and a much-completed surface.

It was also stated that the presence of a stationary drill bit and the number of cycles are significant process parameters for material removal upon Taguchi method analysis. It was concluded that the helical drill bit increases the MRR 2.5 times more than the basic AFF process.

7.2.3.5 Ultrasonic-assisted AFF (UAAFF)

In this technique, the abrasive polymer medium is siphoned with a focal point of the ultrasonically invigorated device. A better finish has been observed when contrasted with the straightforward AFM method. UAAFF was first introduced by Jones and Hull. They combined ultrasonic machining with AFM, which resulted in an improved surface finish by a factor of 10:1 [24].

A two-way horizontal UAAFF on a hollow cylindrical workpiece EN8 steel was investigated. Abrasives come in contact with the workpiece when pushed from one cylinder to another. Meanwhile, the workpiece is treated with high vibrational frequency (5–20 kHz) in the direction perpendicular to the flowing medium, as in Figure 7.7. This results in an inclined collision of abrasive particles, thereby increasing impact force. It was concluded that material removal increments with the increment in the applied frequency of vibration AND extrusion pressure and an increment in surface finish up to 80.12% at an MRR of 7 kg/cm^2 took place [25].

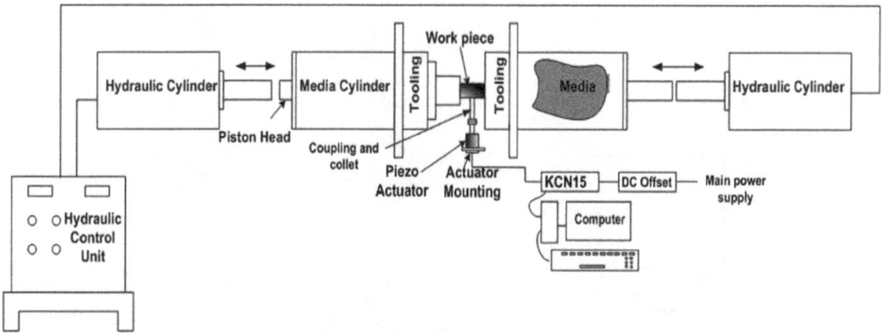

FIGURE 7.7 Schematic of experimental setup of UAAFF [25].

7.2.3.6 Hybrid of electrochemical machining (ECM) & AFM

Hybrid finishing processes are in trend, considering current manufacturing practices. They overcome challenges present in conventional finishing processes owning to the better surface finish of complex, intricate parts and a high MRR.

Experiments have been done upon the hybridization of ECM & AFM and were termed as electrochemical-assisted AFM [26]. The anodic dissolution of the workpiece was caused by the application of polymeric electrolytes. Electrochemical impact and particle movement were both aided by the electrolytic grating medium. In addition to anodic breakdown, mechanical abrasion by abrasive particles also contributed to the material evacuation. Polymeric electrolytes were used, leading to the anodic dissolution of the workpiece. The created electrolytic grating media helped in both the electrochemical impact and the transportation of grating particles. The material evacuation occurred because of rough particles' anodic disintegration and mechanical abrasion.

A process has been examined for incorporating ECM with AFM for the intricate finishing of the interior openings and depressions of the hollow cylindrical workpiece [27]. The process is shown in Figure 7.8.

By forcing a electrolyte-rough laden medium across the gap between the cathode pole and the anode workpiece, the AFM activity component is completed. A yellow metal hollow cylinder was employed as the workpiece. The required nylon fixture was created to enable electrolytic machining with the abrasive cutting activity. Silicon-based polymer (polyborosiloxane) and hydrocarbon gel were combined with NaI salt (a weak alkali) and Al_2O_3 (as abrasives) to create the appropriate electrolytic salt-abrasive packed polymeric medium. Both electrolytic machining and abrasion owing to the abrasive's cutting action take place as the electrolytic salt-abrasive packed medium flows through the annular gap between the fixed cathode rod and the workpiece under a DC voltage.

The AFM activity part is accomplished due to the to and fro expulsion of the electrolyte-rough loaded media through the hole between the cathode pole and the anode workpiece. The workpiece used was a hollow cylinder made up of yellow brass. The necessary nylon fixture was developed to achieve the electrolytic machining along with the abrasive cutting action. The suitable electrolytic salt-abrasive-laden

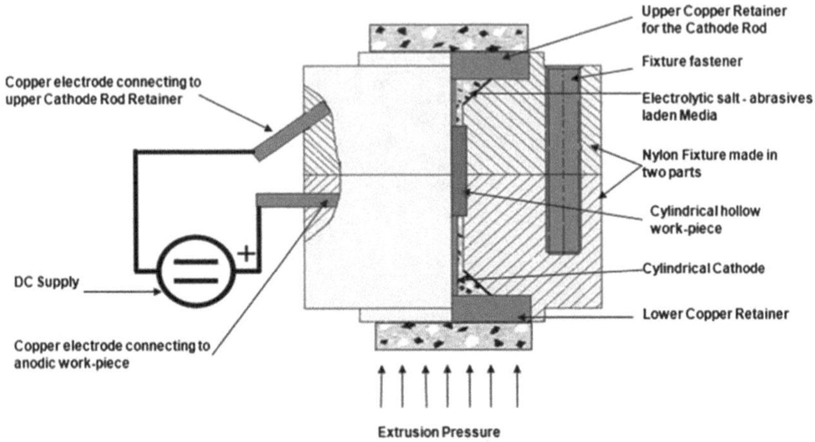

FIGURE 7.8 Hybrid of ECM & AFM [27].

polymeric media was used by adding NaI salt (a weak alkali) and Al_2O_3 (as abrasives) to a blend of a silicon-based polymer (polyborosiloxane) and hydrocarbon gel. As the electrolytic salt-abrasive-laden medium passes through the annular space between the stationary cathode rod and the workpiece under a DC potential, electrolytic machining and abrasion occur due to the abrasive's cutting action. It was concluded that based on Taguchi method parameters, viz. applied extrusion pressure, salt concentration, voltage, electrode gap, the size of the abrasive significantly affected the MRR and surface finish. It was also stated that upon choosing the optimal material removal of 18.02 mg, there was a significant decrease in R_a value from 1.6 μm to a final of 0.84 μm. Thus, the method was influential in the internal surface finishing of cylindrical surfaces of brass as the workpiece material [27].

7.2.3.7 Magnetic abrasive flow finishing (MAFF)

MAFF was developed and a magnetic field was applied around the component undergoing the AFF process. The results showed that as the magnetic field strength was increased, the rate of material removal and the quality of the surface it left behind both improved, and fewer cycles were needed to achieve the same level of removal. It was concluded that upon applying an increasing magnetic field, both MRR and surface finishes increased, and a relatively low number of cycles were required for removing the same amount of material [28].

Carbonyl iron particles (CIPs) and other ferromagnetic particles are sintered with abrasive particles to create a suspension used in abrasive magnetic flow finishing. To bring the CIP closer to the workpiece, a magnetic field is applied from outside the system. In the case of abrasive magnetic flow finishing ferromagnetic particles (such as CIPs) are sintered with abrasive particles, which are then suspended in the medium. An external magnetic field is then applied, which attracts the CIP towards the workpiece. This boosts the MRR by raising the normal force and the concentration of active abrasive particles close to the surface of the workpiece [29]. Altering the strength of the magnetic field has been shown to improve surface quality. Non-ferromagnetic materials may be effectively finished with MAFF. To complete the inner surface of a double-layered tube, Jiang Guo et al. introduced a new method of using rotating-vibrating magnetic abrasive polishing. This increases the density of active abrasive particles in the vicinity near the workpiece's surface, subsequently increasing the normal force and, henceforth, resulting in a better MRR [29].

The surface finish can be optimized by changing the magnitude of the magnetic field applied. MAFF is effective in the case of finishing non-ferromagnetic materials. Jiang Guo et al. presented a novel rotating-vibrating magnetic abrasive polishing method to finish a complex internal surface with a double-layered tube structure. As can be seen in Figure 7.9, a modified lathe setup was used where the workpiece was mounted to the chuck. Magnets were mounted on a pneumatic piston vibrator using a fixture. A finished surface with a roughness value of 0.21 μm (R_a) was obtained [30].

A crossover arrangement of an electro-chemo-magneto abrasive flow machine was used. It expanded the productivity of the customary AFM arrangement by utilizing the attractive electrolytic arrangement in the manufacture [31].

Experiments have been conducted on an inner magnetic abrasive finishing (IMAF) process for non-ferromagnetic complex moulded cylinders. In this cycle, an outer attractive magnetic field was applied, which permitted flexible magnetic

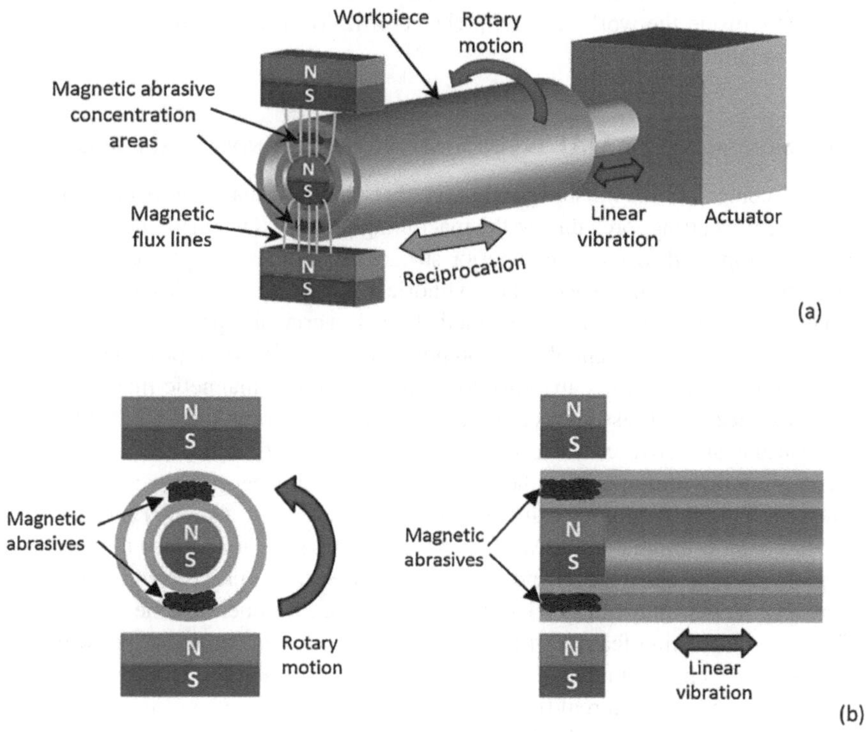

FIGURE 7.9 Magnetic abrasive flow finishing (MAFF) [30].

abrasive brush (FMAB) development inside the cylinder. When the magnetic source is pivoted, a FMAB also starts turning, resulting in the finishing process [32].

7.2.4 Challenges associated with the AFF process for internal cylindrical surfaces

- Basic AFF is a time-consuming process due to the low MRR.
- The MAFF process gives noticeable results only for non-ferromagnetic materials.
- Complex fixtures and arrangements are needed for the hybrid process, like CAAFF, R-AFF, DBAFF.
- The finishing region is uncertain about the active abrasive region.
- The viscosity of the medium used in AFF is challenging to monitor.

7.3 MAF OF INTERNAL CYLINDRICAL SURFACES

The present methodology entails the complete amalgamation of ferromagnetic particles with tiny abrasive particles, namely, Al_2O_3, SiC, CBN, or diamond. Collectively, these entities are denoted as either magnetic or ferromagnetic abrasive particles. A magnetic field is then applied, which attracts ferromagnetic particles close to the surface. As a result, the normal component of magnetic force provides better penetration of abrasives inside the workpiece [33]. Henceforth, MAF can produce components with low roughness value.

7.3.1 Significance of MAF in finishing internal cylindrical surfaces

MAF is a contemporary finishing technique that utilizes a magnetic field to primarily regulate the cutting force during the machining process. This methodology can be effectively employed to produce superior surface smoothness on planar surfaces as well as the inner and outer surfaces of cylindrical workpieces, achieving a finish of a few nanometres. The utilization of magnetic force is a crucial aspect in the creation of FMABs and the advancement of abrasion pressure within the MAF procedure [34].

The technology has the capability to produce non-ferromagnetic materials such as stainless steel and brass, as well as ferromagnetic materials like steel. MAF possesses various attractive characteristics including self-sharpening, self-adaptability, and controllability. Additionally, the finishing tool does not necessitate compensation or dressing. This methodology can be employed across diverse sectors. Empirical evidence has demonstrated that the MAF technique yields superior surface finishes with negligible material elimination and exceptional dimensional precision. Additionally, the process is environmentally friendly and can reduce production time and costs.

The self-sharpening feature of the finishing tool ensures consistent performance throughout the entire finishing process. The self-adaptability feature allows for adjustments to be made in real time to ensure optimal results. Controllability allows for precise control over the finishing process, resulting in a high degree of repeatability. The aforementioned technique can be employed for achieving a surface finish with a roughness of nano levels. This method is particularly suitable for delicate, lightweight, and extremely corrosion-resistant components that are utilized in advanced industries such as automotive, aerospace, medical, and electronics [35, 36]. The utilization of a programmable magnetic field to propel chains of ferromagnetic particles containing abrasive particles that are confined within and between them, in order to conform to the surface of the workpiece to be finished, is a crucial component of this technique. The inherent characteristics of the brush facilitate its ability to access areas that are impervious to conventional tools, such as the interior of curved tubes and pipes [37].

MAF employs a homogenous mechanical combination of abrasive SiC and ferromagnetic iron particles. The pliability of the brush composed of magnetized abrasive particles or a blend of magnetizable particles and abrasive particles is modulated in response to the contours of the workpiece that requires refinement or burnishing. The pliant brush, which exhibits a degree of rigidity as a result of the impact of the magnetic field, applies finishing forces to the surface of the workpiece which wear away the surface, leaving microscopic indentations. As a result, the substance is removed from the brush as it rotates [38].

Micro grinding is a phenomenon that occurs when a workpiece and a flexible brush are subjected to finishing forces and relative motion is provided. This results in the removal of surface material in the form of microscopic chips. At times, the carrier fluid that is blended with the magnetic abrasives possesses electrolytic properties, and this is a deliberate measure to facilitate electrochemical reactions within the operational clearance. In such instances, in addition to abrasive wear, electrochemical wear occurs and adds to the overall material removal [39]. The process of barrel finishing employs the MAF technique, which utilizes permanent magnets to generate a magnetic field. Additionally, it has the potential to be introduced within the aperture of a tube or pipe to refine the inner surface of a non-magnetic stainless-steel cylinder [40, 41].

7.3.2 Parameters to be considered in MAF

There are two types of MAF process parameters: (i) input process parameters, which include things like magnetic flux density, working gap, abrasive grain type, abrasive particle size, abrasive grain relative permeability, free space relative permeability, pole rotation speed, and power input to the pole, and (ii) output process parameters, which include things like finished product size and quality (e.g., surface roughness, weight removal, and accuracy). As can be seen in Figure 7.10, the MAF process is heavily influenced by the abrasive magnetic type, abrasive particles, magnetic device, workpiece material, operating gap, grinding fluid, rotation speed, and axial vibration [42].

Magnetic flux density, working gap, abrasive grain type, abrasive particle size, abrasive grain relative permeability, free space relative permeability, pole rotation speed, and power input to the pole are all significant input elements that exert a considerable impact on the MAF process. The effectiveness of MAF on a flat surface, an inner surface, and a spherical surface is assessed through the use of surface roughness and removal weight as evaluation parameters.

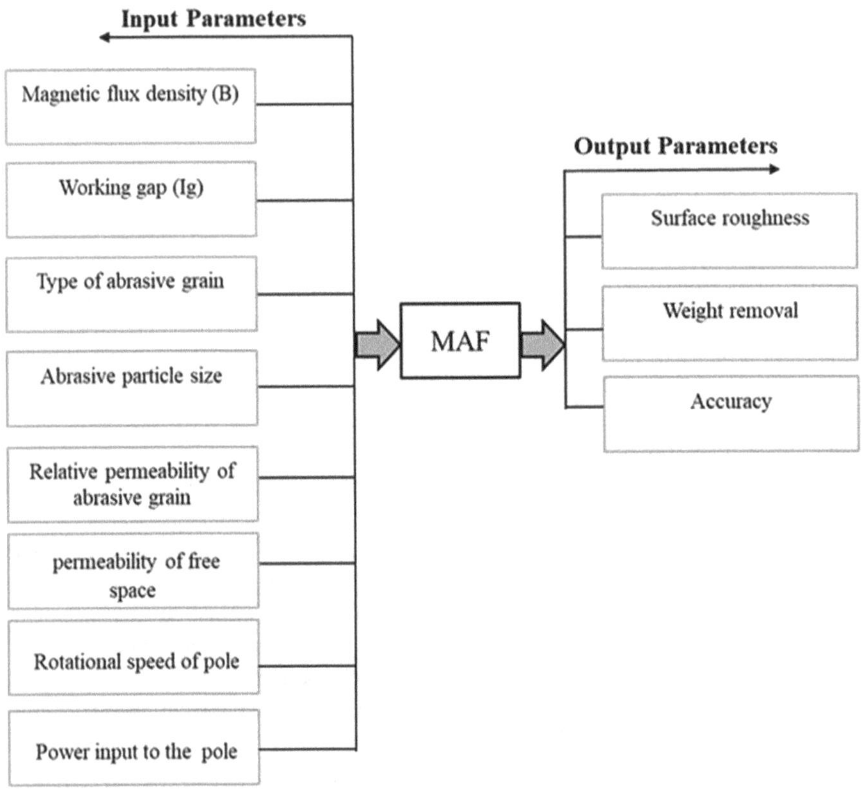

FIGURE 7.10 Schematic diagram of MAF's input and output process parameters.

7.3.3 Variants of MAF for Finishing of Internal Cylindrical Surfaces

7.3.3.1 MAF with rotational pole system

This study investigates the behaviour of an inner surface finishing in a non-ferromagnetic material utilizing a pole rotation mechanism. The workpiece's internal surface undergoes abrasion through the application of magnetic abrasive particles (MAPs) that are introduced into the workpiece and subsequently drawn towards the magnetic field [43]. The magnetic field at the finishing zone is rotated by the spinning of the small permanent magnet poles around the workpiece. The deleterious effects of the rough behaviour exhibited by the inner surface of the tube have a notable impact on the finishing characteristics. This behaviour can be categorized into four distinct stages, as visually depicted in Figure 7.11(a). The described phenomenon involves the confinement of all abrasive particles through magnetic means within the finishing area, allowing for their seamless rotation along the inner surface in response to the movement of the magnetic field. This facilitates the enhancement of surface roughness.

Figure 7.11(b) illustrates the rotational movement of the abrasive specimens against the inner surface, while simultaneously experiencing self-displacement within the

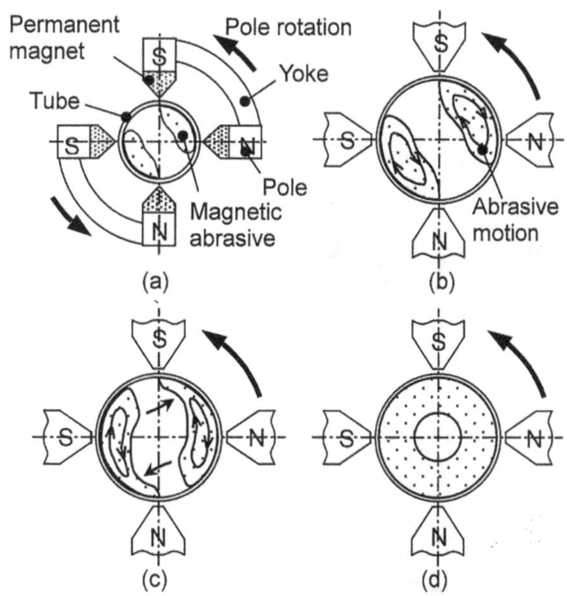

FIGURE 7.11 Two-dimensional schematic of abrasive behaviour during the process [43].

abrasive mass. The phenomenon known as "self-displacement" pertains to a state wherein the abrasive particles relocate within the abrasive matrix. The self-displacement of the abrasive's cutting edges may lead to an increase in material removal. The observed behaviour depicted in Figure 7.11(c) is attributed to a decrease in the magnetic attraction exerted on the abrasive material. Under these conditions, the magnetic field exhibits a limited capacity to retain a fraction of the abrasive mass within the finishing zone. The distribution of the abrasive material occurs in a random manner, causing it to impact the surface area and subsequently induce cutting action [43].

7.3.3.2 Ultrasonic vibration-assisted MAF process

The illustration presented in Figure 7.12 showcases the ultrasonic-assisted magnetic abrasive finishing (UAMAF) technique, which is a hybrid finishing process that utilizes ultrasonic vibrations within the finishing zone of a MAF process to achieve improved surface topology within a reasonable timeframe. Empirical evidence indicates that UAMAF yields superior results compared to MAF when utilizing identical finishing parameters [44]. The UAMAF configuration is depicted in both a schematic and an experimental system in Figures 7.12(a) and 7.12(b), respectively. The experimental setup consisted of a specially designed workpiece fixture and an ultrasonic vibration generator. The arrangement of the four poles of the electromagnet followed a north-south orientation, resulting in the generation of magnetic fields that were concentric in nature.

A comparative analysis has been conducted to evaluate the efficacy of MAF and UAMAF in the context of internal surface finishing of hollow ceramic pipes, as shown in Figure 7.12(c) The study presents a comparative assessment of the surface

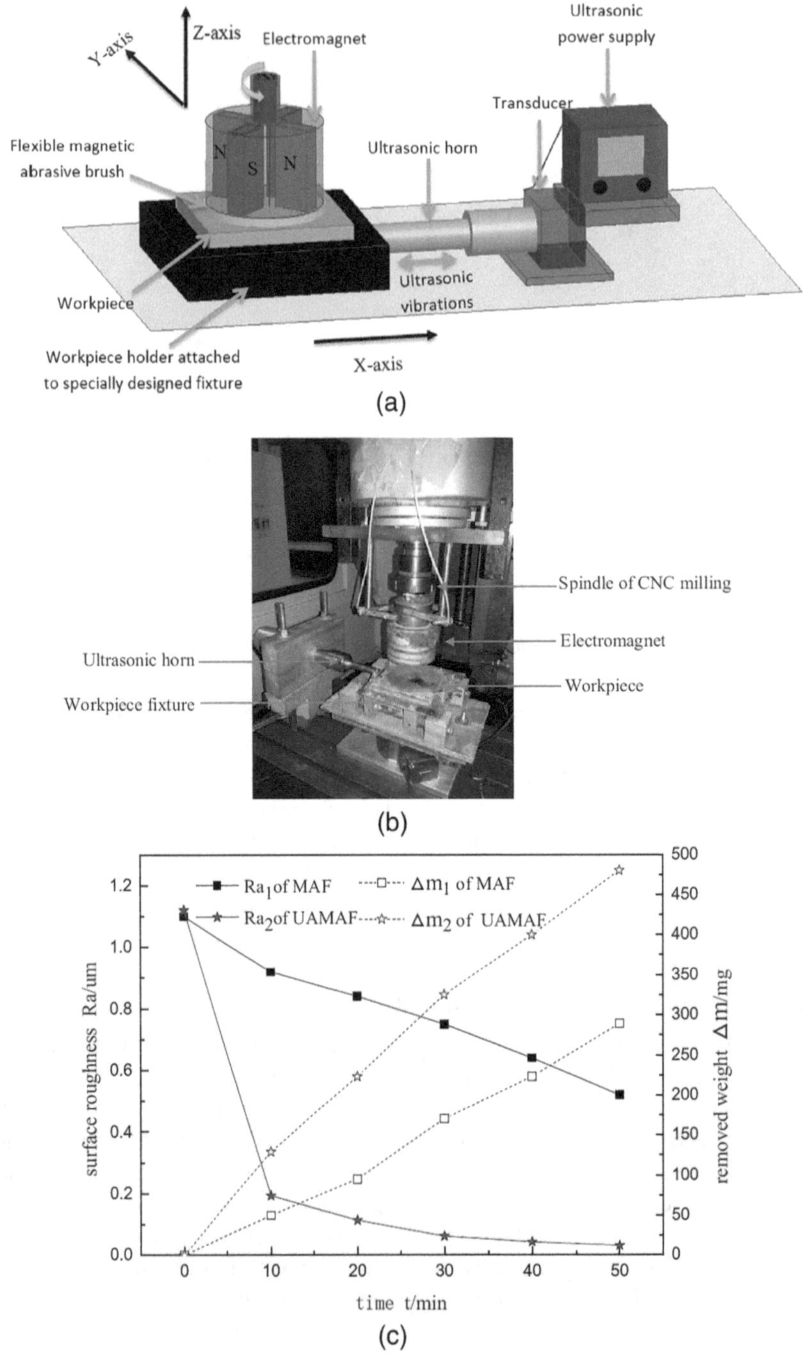

FIGURE 7.12 (a) Schematic diagram of UAMAF process. (b) Actual experimental setup of UAMAF process and (c) comparative analysis of surface roughness and material removed with finishing time of MAF and UAMAF [44, 45].

roughness and material removal capabilities of both processes [45]. The inferiority of MAF in terms of surface roughness and material removal compared to UAMAF is evident. Upon completion of a 50-minute process, the surface roughness of the material known as MAF exhibited a decrease from 1.1 to 0.55 μm, while the amount of material removed was 295 mg. The UAMAF exhibits a decrease in surface roughness from 1.12 to 0.03 μm, accompanied by a material removal of 485 mg. The UAMAF serves to augment the frictional forces between the MAPs and the workpiece, thereby facilitating an increase in the kinetic energy of the MAPs. Consequently, the process of material removal experiences a significant enhancement. Conversely, the application of high-frequency vibration yields a positive impact on trajectory interference by reducing the size of non-removal traces, resulting in a smoother trajectory and a noticeable enhancement in surface quality [45].

7.3.3.3 MAF with a multiple pole tip system

A proposed MAF process utilizing a multiple pole tip system has the magnetic attributes of a metastable austenitic stainless-steel tool and can be modified through the process of heat treatment. By subjecting the tool to multiple partial heat treatments, it is possible to create alternating regions with magnetic and non-magnetic properties. The utilization of this tool within a capillary tube, in conjunction with the magnetic abrasive, results in the creation of numerous finishing regions. This occurs due to the magnetic attraction that occurs between the magnetized and not magnetized regions on the tool, which are present at the border. The passage of the magnetic abrasive deep into the tube is facilitated, allowing for the simultaneous finishing of multiple regions [32]. As shown in Figures 7.13(a) and 7.13(b) the proposed approach involves the utilization of a multiple pole tip system.

This study involves the incorporation of a pair of pole tips into a single pole tip system, which is then interrelated to produce magnetic fields within two distinct finishing areas. The finishing area is expanded by incorporating a magnetic abrasive into two specific regions of the tube, which exerts pressure against the inner surface of the tube. Similar to the single pole tip system, the rotation of the tube facilitates the movement of the magnetic abrasive in relation to the inner surface of the tube, which leads to a highly effective surface finishing process [32].

In practical applications, it has been shown that certain mixed-type magnetic abrasives tend to cling to the surface of a tube. This adhesion occurs due to the friction between the mixed-type magnetic abrasive and the tube surface, which surpasses the magnetic force that would otherwise induce the mixed-type magnetic abrasive to follow the rotation of the pole tip. As depicted in Figure 7.13(b), the area corresponding to the chuck-end edge of the pole tip should possess a minimal quantity of magnetic abrasive.

Consequently, it can be observed that an increase in the depth of the finishing region within the tube leads to a corresponding increase in surface roughness. This can be attributed to the insufficient presence of magnetic abrasive in the finishing process. Consequently, the initial penetration of MAPs into the tube and the ability to supply a mixture of different types of MAPs are crucial factors in achieving the desired outcome in terms of finishing performance. The design of the alternating magnetic property and the spacing between the magnetic and non-magnetic zones must be carefully considered to address these concerns.

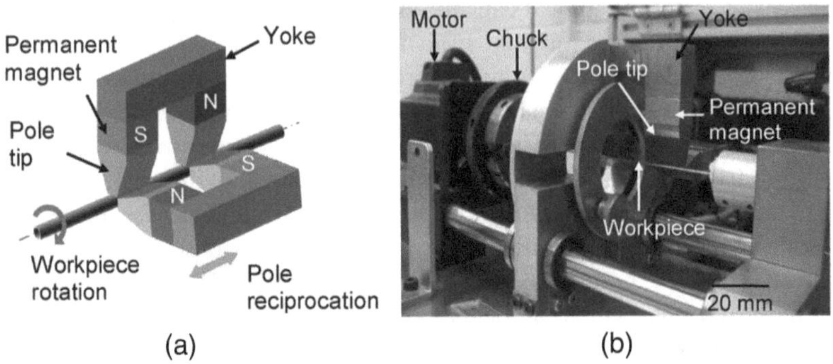

FIGURE 7.13 (a) Schematic diagram of multiple pole tip system. (b) Actual experimental setup of multiple pole tip system [32].

7.3.3.4 Finishing internal circular tubes using magnetic abrasive jet machining

The proposed method involves the utilization of a precise internal finishing technique that employs a working fluid combined with magnetic abrasives. This mixture is then directed into the internal surface of the tube, while magnetic poles are positioned on the exterior surface of the tube. This study aimed to develop a novel finishing technique known as the magnetic abrasive jet machining system [46].

Figure 7.14 depicts a schematic representation of the magnetic abrasive jet machining system (MAMS), which comprises several components including a fluid supply system, a nozzle, electromagnet parts, a power source capable of delivering a voltage range of 0 to 100 V with a maximum current of 5 amp. a mixing chamber, discharge parts, and an abrasive tank. The working fluid in this system is compressed air, with a maximum pressure of 4–5 atm. The flow rate of the compressed air is regulated by a valve, with a maximum flow rate of 250 ml min^{-1}. The automatic supply of magnetic abrasive is facilitated through the utilization of jet fluid. Consequently, the magnetic abrasive and the working fluid are combined. A heterogeneous fluid is injected into the interior of the workpiece by a nozzle, afterwards flowing into the discharge pipe at a low pressure and velocity. The abrasive that has been expelled from the tank is subsequently recycled by an abrasive feed pipe [46].

Figure 7.15 depicts the comprehensive schematic diagram of the machining principle. The system is capable of completing the task exclusively through the jet process, as depicted in Figure 7.15(a) However, Figure 7.15(b) illustrates a modification to the machining principle by incorporating electromagnets into the jet process. Electromagnets are incorporated to enhance the machining force and improve the distribution of magnetic abrasive on the internal surface, thereby enabling more efficient precision finishing. The primary method employed for basic machining is the jet process, wherein the electromagnets play a crucial role in enhancing the efficiency of the machining process [46].

7.3.3.5 Chemical-assisted magnetic abrasive finishing (CMAF) process

The utilization of titanium alloys, tungsten, and other composites has become prevalent in contemporary industry due to their superior strength, hardness, wear

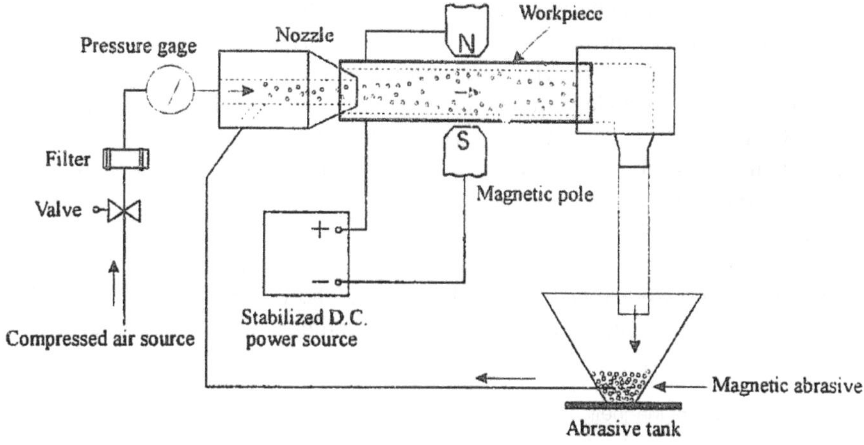

FIGURE 7.14 Schematic of magnetic abrasive jet machining (MAJM) [46].

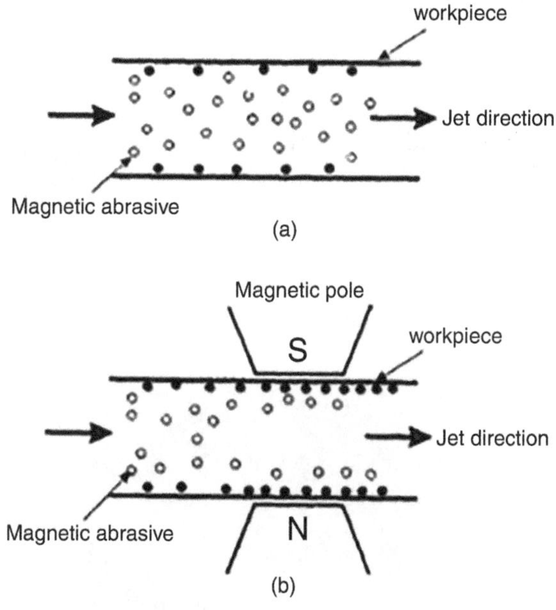

FIGURE 7.15 Schematic diagram of machining principle (a) with magnet; (b) without magnet [46].

resistance, and toughness in comparison to conventional materials. Yet, due to these features, they are difficult to comprehend. MAF and chemical mechanical polishing (CMP) are alternative techniques that can be employed for the finishing of these materials, albeit with a lower level of productivity. MAF is a material removal technique that utilizes MAPs and a magnetic field to eliminate microchips from a workpiece [47, 48]. MAF and CMP are both widely used because they provide a variety of benefits. Their low efficiency and MRR are disadvantages, especially when working with tough substances. Thus, businesses see the need to create a combined finishing process that improves efficiency without sacrificing surface quality.

Figure 7.16 depicts the CMAF process for simultaneous interior and exterior surface finishing. The workpiece (tube) is first chemically treated with an appropriate etchant. On the work surface, an appropriate etchant is applied and held at a particular high temperature for a set period of time. By reacting chemically with the work surface, the etchant softens the surface's molecular layer, allowing for its removal via an MAF technique [49].

The MAF procedure involves the placement of two permanent magnets around a chemically treated workpiece, specifically a tube. The vertical cross-sectional profiles (North-South orientation) of both magnets exhibit a precise congruence with

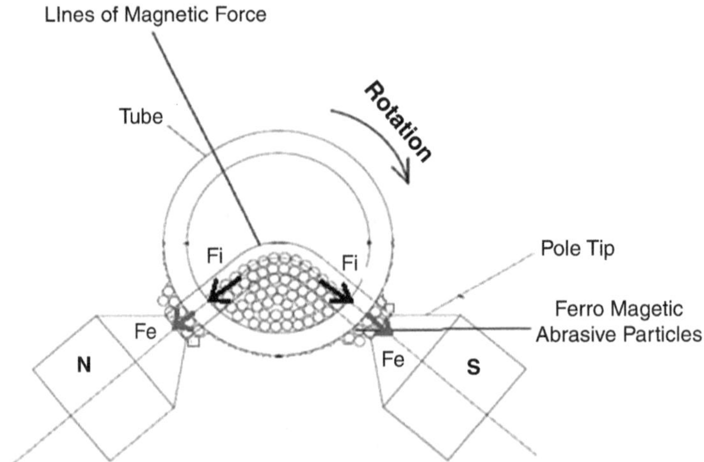

(a)

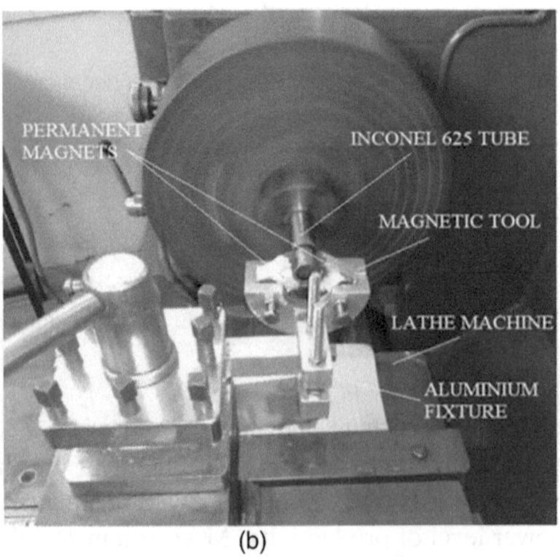

(b)

FIGURE 7.16 (a) Schematic representation of CMAF process. (b) Actual setup of CMAF process [49].

the exterior wall of the surrounding tube. The tube is equipped with ferromagnetic abrasive particles that are present internally as well as externally. The formation of an FMAB occurs through the attraction of ferromagnetic abrasive particles to the magnetic field. The magnetic force F_i generated within the tube draws the MAPs inwards. Magnetic particles are coaxed outwards by an extraterrestrial force known as F_e, which is created outside the tube [49]. Figure 7.16(b) depicts the experimental setup used to finish both the interior and outside surfaces of Inconel 625 tubes at the same time.

To simultaneously clean the internal and external surfaces, a native magnetic tool has been developed. Figure 7.16(b) illustrates how it is made up of two permanent neodymium iron boron (NdFeB) magnets mounted on an aluminium frame, each measuring 35 × 35 × 25 mm. Both magnets' upper pole profiles are perfectly aligned with the tube's exterior surface. The magnet poles are covered with nonwoven fabric coated with PTFE (polytetrafluoroethylene) tape. An experimental setup involves the attachment of a magnetic tool onto the tool post of a multispeed precision lathe.

7.3.3.6 MAF with a stationary pole system

A novel method of internal MAF has been developed to achieve superior surface finishing of workpieces employed in crucial applications such as semiconductor manufacturing facilities. The methodology employed utilizes an uninterrupted magnetic field to administer a finishing force to magnetic abrasives. Workpieces with rotation-symmetrical shapes and tiny dimensions, such as straight tubes, gas bombs, reducers, and so on, are rotated at high speeds using a stationary pole system [50].

Figure 7.17 depicts the concept of an internal MAF process utilizing a stationary pole system. The magnetic field is responsible for consolidating the magnetic abrasive that is introduced into the tube at the finishing zone, thereby exerting the finishing force on the inner surface of the tube [50]. In practice, electromagnetic coils or permanent magnets can be used to generate magnetic fields [50, 52]. When the tube

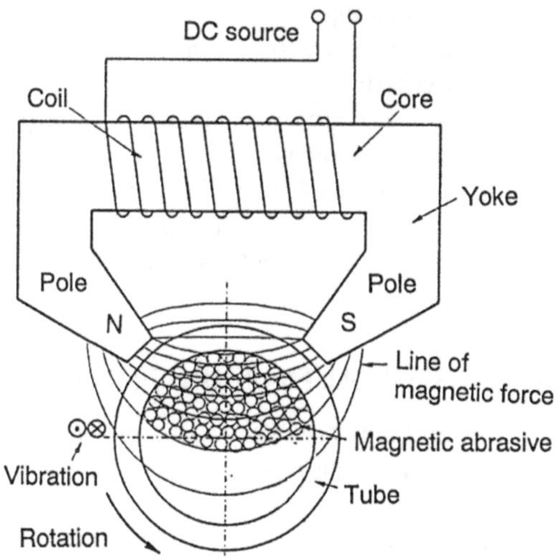

FIGURE 7.17 Schematic internal MAF process using stationary pole system [51].

rotates fast, the material removed by the inner surface of the tube is smoothed by the relative motion between the magnetic abrasive and the inner wall of the tube.

7.3.4 Challenges Associated with the MAF Process for Internal Cylindrical Surfaces

Despite the potential advantages of MAF procedures, there exist several limitations:

- Geometries like a workpiece's wrinkled surface are still difficult to finish effectively with the MAF technique [53].
- As a result of its ineffectiveness on tough and rigid materials, MAF has a low MRR [54].
- The MAF technology was devised for the purpose of expediting the process of interior finishing and cleansing of pliable capillary tubes. The efficacy of non-conventional techniques in eliminating burrs from the interior surface of a tube diminishes with a reduction in its diameter [40].
- The research conducted on the impact of rotational motion on the MRR of a flat workpiece has revealed that a low MRR is a significant constraint in almost all of these procedures [41].
- MAF media has the ability to impregnate the workpiece's surface [41].
- The interiors of flexible capillary tubes were expeditiously completed and cleaned through the utilization of MAF. As the pipe diameter decreases, it becomes more and more challenging to remove burrs completely from the inner surface using unconventional techniques.
- By applying three-phase AC current to three coil sets at 120 degree intervals in three different directions, a revolving magnetic field can be produced, which is ideal for interior surface finishing of cylinder-shaped workpieces. The present study involved conducting experiments on mild steel, hardened steel, and brass specimens utilizing MAPs. These particles were produced by amalgamating Al_2O_3 and iron particles, and a straight oil type grinding fluid was employed for the purpose. After conducting tests, it was discovered that utilizing a revolving magnetic field increases stock removal but lowers surface quality when compared to using a static magnetic field. The magnetic field intensity of non-ferromagnetic substances is relatively low due to the absence of magnetization in the constituent components. Consequently, the process of stock removal is significantly restricted, and the recommended equipment is not suitable for materials that are non-ferromagnetic, as mentioned [55].

7.4 MAGNETORHEOLOGICAL FINISHING (MRF)

Magnetorheological polishing fluid (MRPF)-assisted finishing techniques represent a category of finishing procedures that provide enhanced process control, enabling the attainment of fine tolerances and preservation of surface topography integrity. The utilization of magnetorheological polishing fluid MRPF has been observed in various contemporary magnetorheologicalpolishing fluid MRPF-assisted finishing techniques, wherein it serves the purpose of externally controlling the finishing forces exerted on abrasive particles [56]. Several MR finishing techniques include MRH, rotational magnetorheological honing (R-MRH), MRAFF, and R-MRAFF.

The utilization of MR fluid-based finishing (MRFF) involves the application of a liquid carrier that contains a combination of magnetic particles and abrasive particles within its composition, commonly referred to as MR fluid. Chemicals and surfactants, commonly referred to as additives, are employed to prolong the lifespan of the MR fluid and enhance the rate of particle elimination [57]. MRF is a high-precision methodology that can attain surface precision of a 30-nm peak to valley and micro roughness of less than 10 Å. The utilization of MRPF is applicable to a diverse array of materials, encompassing optical glasses, hard crystals, and various metallic surfaces, as indicated by source [58].

7.4.1 Significance of MR finishing in finishing of internal cylindrical surfaces

Magnetic-assisted finishing techniques, such as MAF, are being developed to provide a high level of workpiece surface finishing with regulated finishing pressures. Using magnetorheological polishing fluid (MRPF), magnetic-assisted finishing procedures are able to complete a broad variety of workpiece forms at the nano-scale level while maintaining high surface integrity and eliminating defects including microcracks, heat-affected zones, and residual stress. An illustration of this type of finishing process is magnetorheological finishing (MRF). MR polishing fluids are utilized for these processes; they are a smart and magnetically regulated blend of abrasive particles and ferromagnetic particles and are mixed with the base fluid which is a grease and oil mixture [57].

Magnetorheological finishing (MRF) is an exemplar of this category of finishing techniques. The aforementioned processes employ MR polishing fluids, which consist of a sophisticated combination of abrasive and ferromagnetic particles, blended with a base fluid comprising a mixture of grease and oil [59]. Upon exposure to a magnetic field, ferromagnetic iron particles undergo magnetization and exhibit a chain-like morphology at the microscopic level. The abrasive particles, which are confined within the chains of magnetic iron particles, offer ample support as evidenced by previous studies [53, 54].

MRAFF and R-MRAFF are employed for the purpose of refining the internal cylindrical surfaces of through holes that are predominantly composed of non-ferromagnetic substances [57].

7.4.2 Parameters to be considered in MRF

The MRF technique is a finishing process that relies on fluid-based mechanisms. The process parameters are predominantly determined by the components and concentrations of the MR fluid. Figure 7.18 displays several crucial parameters of the MRF procedure. In order to achieve optimal performance from the MRF process, which is a challenging task, it is imperative to exercise stringent control over the process parameters. The selection of process parameters and their values is influenced by the composition, hardness, and initial surface roughness of the workpiece to be processed.

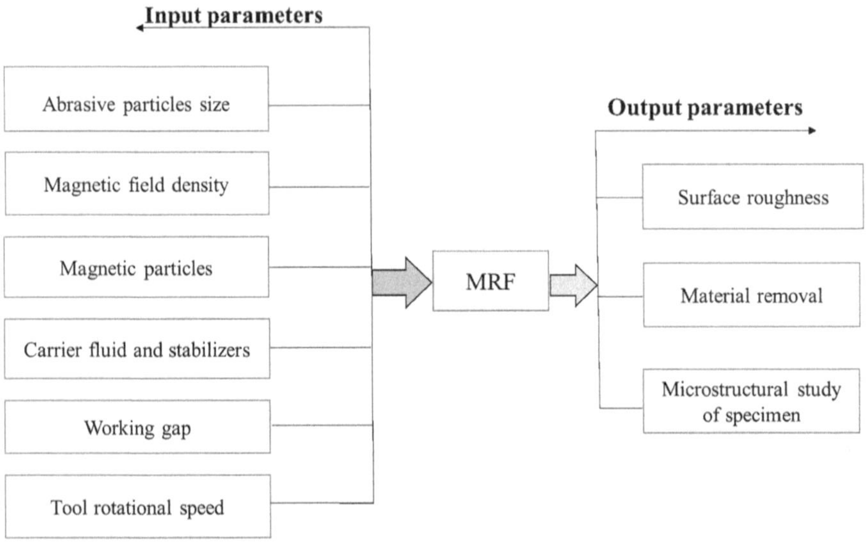

FIGURE 7.18 Input and output process parameters to be considered in MRF.

In comparison to other process factors, the magnetic field and the size and concentration of CIPs have a more significant influence on the MRF process, as the absence of either of these two factors results in the inability of the MRPF to function [57]. In order to achieve the desired surface roughness and MRR, it is imperative to select the optimal level of said parameters.

The MR fluid's exceedingly high yield stress renders it inappropriate for the finishing of delicate glass or silicon-based materials, owing to its forceful interaction with the workpiece, leading to the occurrence of significant micro and nano scratches. Therefore, it is desirable to have a moderate yield stress in order to achieve the ultimate low R_a and high MRR. The selection of a process parameter is a critical decision, with the type of workpiece material (e.g., metal, optics such as glass or lenses, ceramic, etc.) and the initial surface roughness being the primary factors to consider. When dealing with optics, a material that is characterized by its softness and relatively smooth surface, it is imperative that the composition of the MR fluid used does not cause excessive abrasion to the surface of the workpiece, as mentioned [57].

7.4.3 Variants of MRF for Finishing of Internal Cylindrical Surfaces

7.4.3.1 Magnetorheological honing

The inner surfaces of cylinders made of ferromagnetic and non-ferromagnetic materials may now be finished to a high standard using novel MR fluid-based finishing processes [60, 61]. The suggested MRH method nano-finishes the inner surface of cylinders of varying diameters using four flat permanent magnet strips. Further improvements to the recently existing MRH method have been made in this study to improve its finishing performance [61].

The models represent two identically sized MRH tools (one with flat permanent magnet strips and the other with curved permanent magnet strips). Both MRH tools include four permanent magnet strips along the outside edge of the finishing tool. Models of the MRH process utilizing flat and curved magnet strips are shown in Figures 7.19(a) and 7.19(b). The four magnetic strips form a right angle with respect to the finishing tools' central axis [62].

The models depict two MRH tools of equal dimensions, one equipped with flat permanent magnet strips and the other one with arched permanent magnet strips. The finishing tool of both MRH tools is equipped with four permanent magnet strips positioned along its outer edge. Figures 7.19(a) and 7.19(b) depict the various models of the MRH process that employ flat and curve-shaped magnet strips [62]. The orientation of the four magnetic strips is perpendicular to the central axis of the finishing tools.

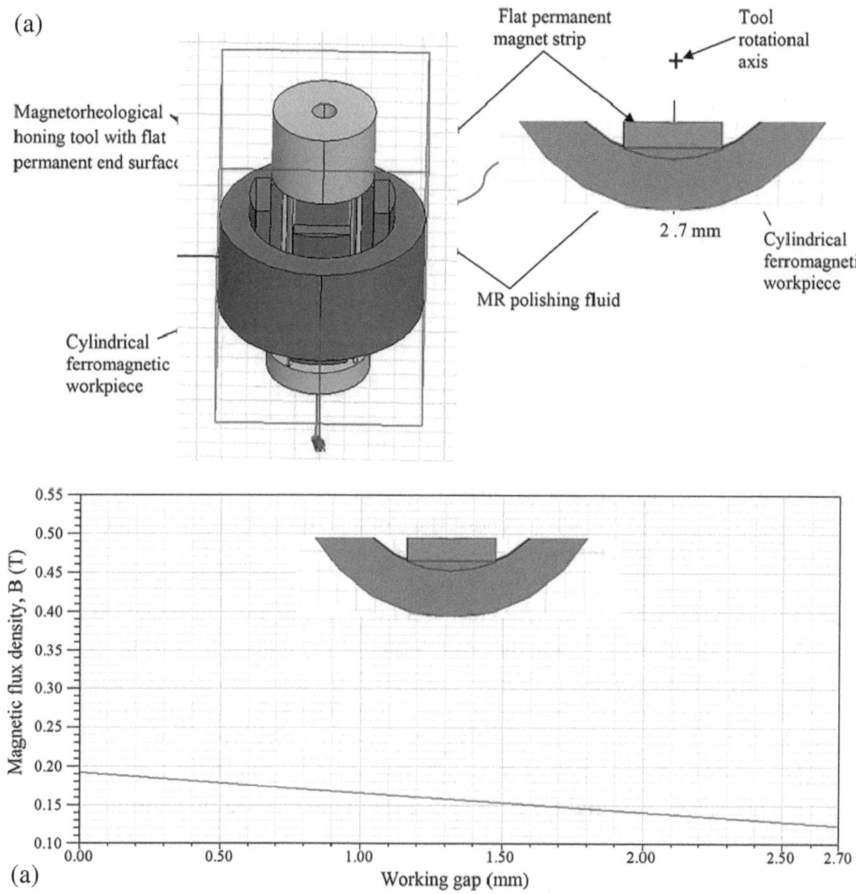

FIGURE 7.19 a Magnetic models of MRH tool and magnetic flux density in the MR polishing fluid area from the (a) flat and (b) end surface of a permanent magnet strip to the inner surface of a cylindrical ferromagnetic workpiece [62].

(b)

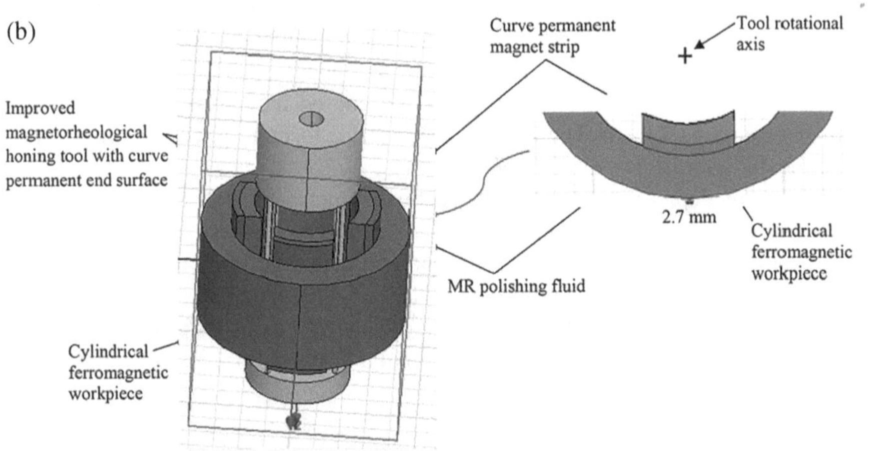

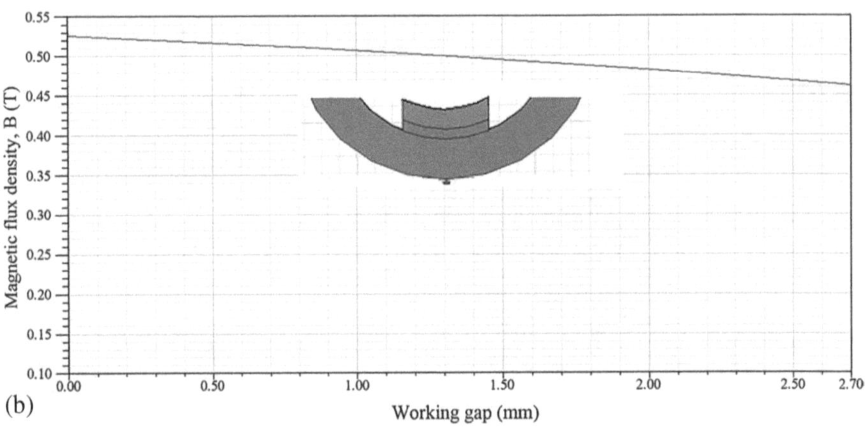

(b)

FIGURE 7.19 b Magnetic models of MRH tool and magnetic flux density in the MR polishing fluid area from the (a) flat and (b) end surface of a permanent magnet strip to the inner surface of a cylindrical ferromagnetic workpiece [62].

The functional behaviour of a tool utilized for finishing the inner surface of a cylinder may vary depending on whether its magnetic core end surface is flat or curved. The reason behind this phenomenon is attributed to variations in the magnetic flux density across the surface of the finishing tool [62]. The MR polishing fluid achieves uniformity in bonding force upon finishing the surface of the tool in the working gap, resulting in uniformly distributed magnetic flux density over the finishing tool surface. This process is essential for completing the inner surface of a hollow cylinder-shaped workpiece.

7.4.3.2 Magnetorheological abrasive flow finishing

Magnetorheological (MR) fluid, comprising magnetic CIPs, non- MAPs, water or oil, and diverse additives, was developed for the purpose of MRAFF [63, 64]. The magnetic field used keeps abrasive particles together in a vast network of CIPs.

As a result, the cutting motion is enhanced, producing greater MRR and surface quality.

The MRAFF technique entails the iterative extrusion of a magnetically stiffened slug of MRPF through the channel that is created by the workpiece and fixture [63]. The application of a magnetic field onto a workpiece results in selective abrasion of only the area that is exposed to the magnetic field, while the remaining surface remains unaffected.

The process mechanism is depicted in Figures 7.20(a) and 7.20(b). The development of a refined finishing apparatus has the potential to enhance the utilitarian properties of said ferromagnetic cylindrical constituents through increased regulation. A correlation has been observed between the decrease in surface roughness values and a reduction in wear and energy consumption exhibited by the mating components [61]. In contrast to conventional finishing tools, this tool has the potential to enhance the longevity and upkeep of said components.

In order to solve the difficulties associated with the thorough finishing of ferromagnetic workpiece surfaces, a technique based on MR fluids has been developed. The finishing tool and electromagnet in this method were contained within the cylinder of the ferromagnetic workpiece. This method involved housing both the electromagnet and the finishing tool within the cylinder of the ferromagnetic material.

The magnitude of the material removed from the crests of the workpiece's surface by abrasive grains is ascertained by the bonding potency conferred by the field-induced configuration of the MR polishing fluid and the extrusion pressure

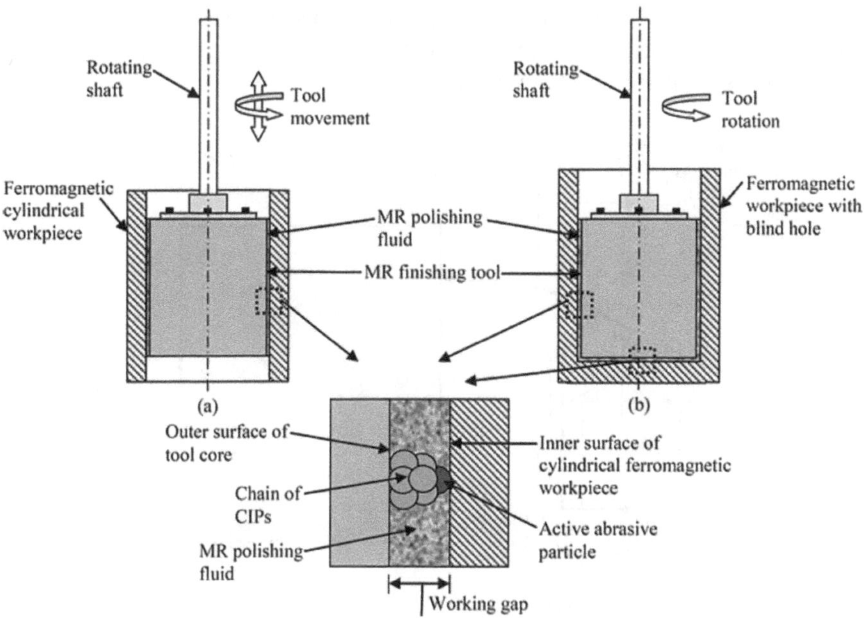

FIGURE 7.20 Schematic of MR fluid-based finishing tool during finishing of the internal ferromagnetic (a) cylindrical work piece and (b) blind hole workpiece [61].

dispensed through the piston. The regulation of peak abrasion can be achieved by adjusting the magnetic field strength [65].

7.4.3.3 Rotational magnetorheological abrasive flow finishing

The R-MRAFF method is a distinctive finishing technique that has been suggested to improve the finishing capabilities of the MRAFF process. The process of polishing involves utilizing a magnetic field and hydraulic unit to apply a circular and reciprocating motion to the medium. This is depicted in Figures 7.21(a) and 7.21(b). The efficacy of the process was validated through the polishing of stainless steel and brass plates [66].

The R-MRAFF technique employs a hydraulic system to operate two opposing pistons within the medium cylinders, thereby facilitating the extrusion of magnetorheological polishing (MRP) medium onto the surface of the workpiece, and creating a vertical motion along the tube axis. Figure 7.22, depicts the simultaneous rotation of permanent magnets (NdFeB) that surround the workpiece fixture and the polishing media (magnetic in nature). When controlled precisely, these two motions may provide a surface finish that is uniformly smooth and mirror-like over the whole range [67].

Forces are exerted on a particle of abrasive material by the R-MRAFF procedure. The magnetic force F_m acting on a CIP in a magnetic field causes the abrasives contained by the CIPs to be forced onto the surface of workpiece [66].

$$F_m = m\mu_0 \chi_m \mathrm{H}\nabla\mathrm{H} \tag{7.1}$$

The equation for the magnetic force acting on a CIP is expressed in Eq. (7.1) where m represents the mass of the CIP particle, χ_m denotes its mass susceptibility, H is the magnetic field intensity, ∇H represents the gradient of the magnetic field intensity, and μ_0 denotes the permeability in free space.

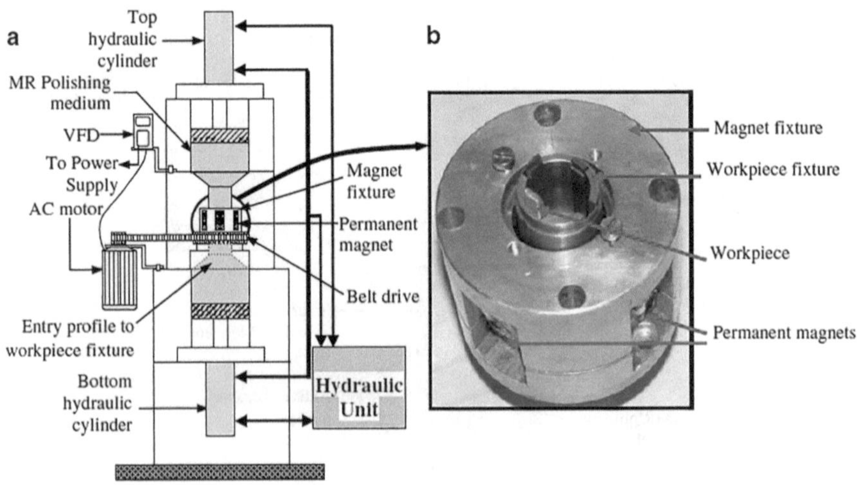

FIGURE 7.21 (a) Schematic diagram of R-MRAFF setup and (b) magnet fixture along with workpiece fixture [66].

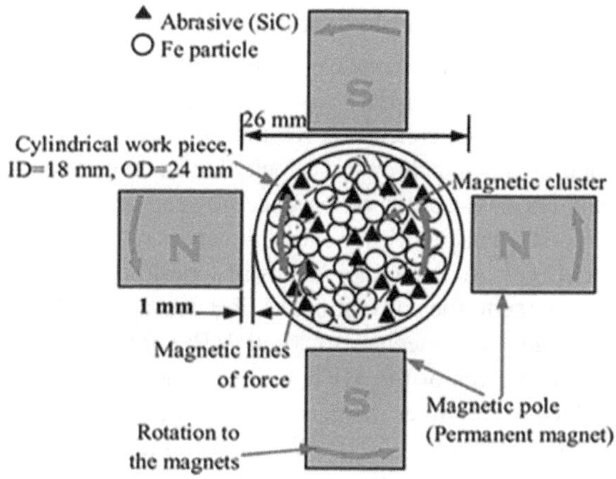

FIGURE 7.22 Schematic diagram showing rotation of magnets and medium during the R-MRAFF process [67].

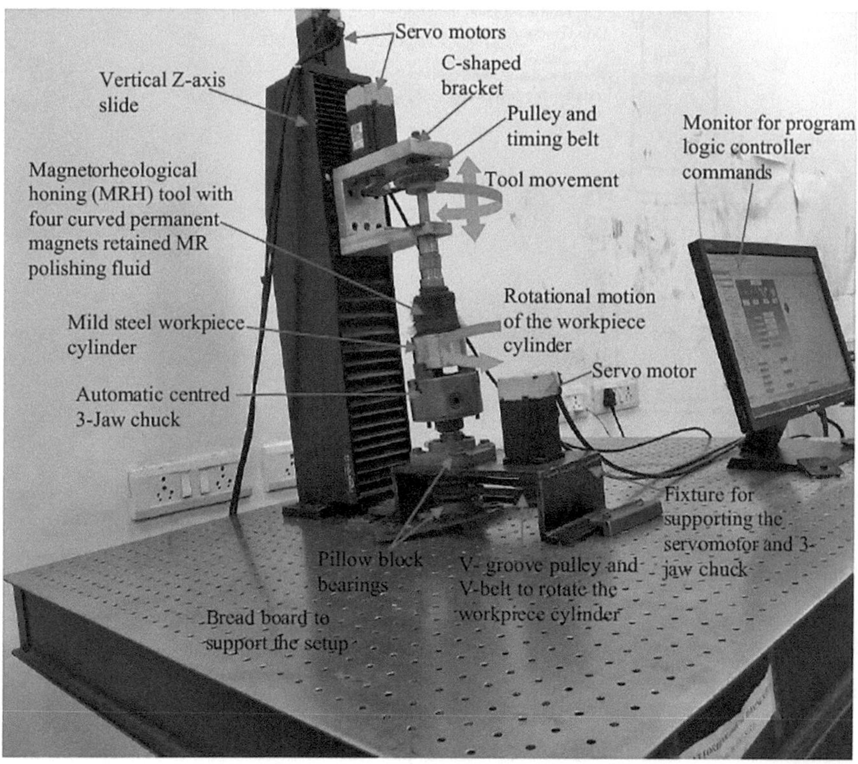

FIGURE 7.23 Actual experimental setup of R-MRH [68].

7.3.3.4 Rotational magnetorheological honing

The current MRH technique may be made even more efficient. To speed up the completion time of conventional MRH, R-MRH was developed. As depicted in Figure 7.23, the R-MRH technique involves the concurrent rotation of the cylinder and the MRH tool, with the former rotating at a synchronous velocity but in an opposing direction to the latter. An additional shear force can be produced by the active abrasives due to the increased relative speed when the cylindrical workpiece is rotated in the opposite direction of the MRH tool. This methodology results in superior surface finishing within a shorter duration compared to the conventional MRH approach [68].

Figure 7.24 illustrates the mechanism of R-MRH for surface finishing which describes that the end surface of the MRH tool generates a magnetic field (F_m) that normalizes the surface of the inner cylinder through the use of confined active abrasives containing a chain structure of CIPs. When utilizing an active abrasive particle for the purpose of polishing a workpiece, it applies a magnetic force onto the surface

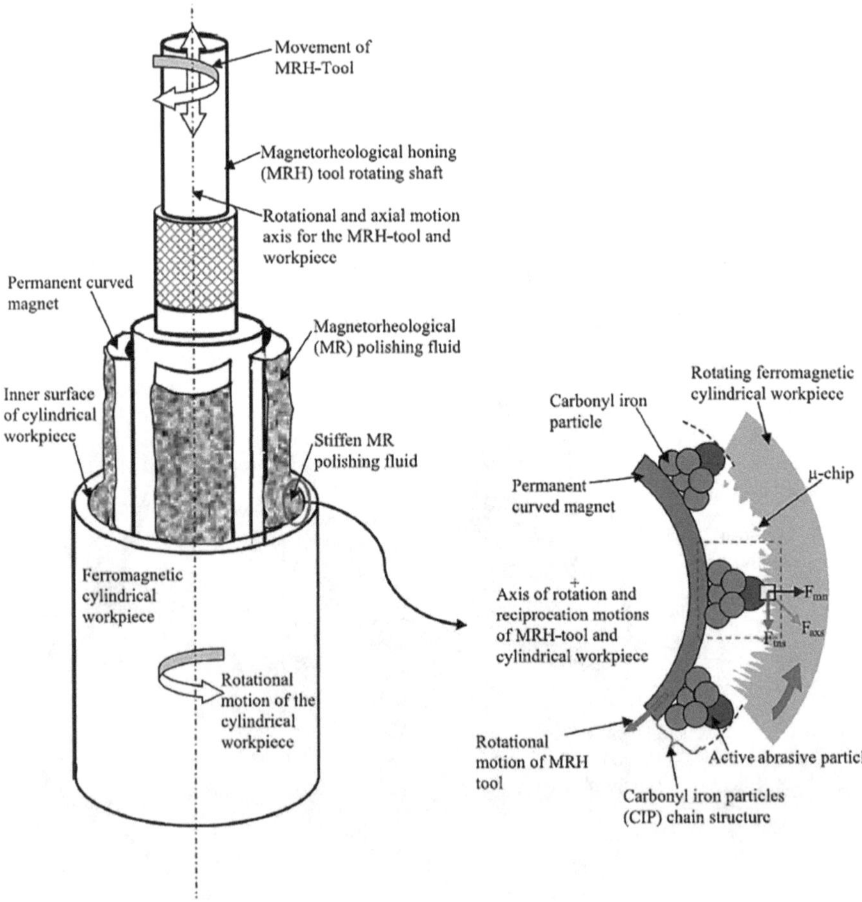

FIGURE 7.24 Schematic representation of the finishing mechanism in the R-MRH process [68].

that is equivalent to the magnetic force encountered by a CIP [68]. One can determine the magnetic force (F_m) produced by a CIP by utilizing Eq. (7.2).

$$F_m(x) = \frac{mx_m}{\mu_o} B(x) \frac{\mathrm{d}B(x)}{dx} \tag{7.2}$$

where F_m is a function of the horizontal distance between the end surface of the MRH tool and the inner surface of the cylindrical workpiece, i.e., the working gap, x. CIPs have different magnetic susceptibilities depending on their position due to the magnetization of the materials and the magnetic field created there [65].

7.4.4 Challenges associated with the MRF process for internal cylindrical surfaces

- The MRP fluid used in MR procedures is more expensive since it must be conditioned on a regular basis. As a result, the entire lead time increases.
- Because the MR process is a complicated interplay of electromagnetic coils, MR polishing fluid, and workpiece material characteristics and form, it's impossible to anticipate how newer materials will respond.
- Due to the high current flow, an electromagnetic coil utilized in the MR process heats up. It causes a loss in efficiency, a rise in coil resistance, and a change in the characteristics of the MR fluid.

7.5 CONCLUSION

- AFF is classified as a non-conventional finishing method that involves the utilization of a semi-solid medium composed of a visco-elastic polymer and abrasive particles, which are combined in a predetermined ratio. The media undergoes extrusion under pressure, either through or across its surface, in order to achieve the desired finish.
- The MAF process is efficacious in enhancing the surface accuracy and dimensional precision of diverse workpiece geometries such as internal finishing of hollow cylindrical workpiece, capillary tubes, and concave-shaped surfaces. The MAF technique has the capability to be employed with both bonded and unbonded variants of magnetic abrasives. The utilization of unbounded magnetic abrasives in MAF has been found to result in increased MRR, but the use of bonded magnetic abrasives has been observed to offer superior surface quality.
- For nano-finishing operation of internal surface of cylindrical components using MR finishing processes very few experimental setups have been designed and fabricated till now for very shorter length of the cylindrical components using different variants of the MR finishing process. There is a lot of scope for the design and fabrication of MR finishing setups for longer length of cylindrical components which have applications in manufacturing and automotive industries such as injection barrels, cylindrical moulds, hydraulic piston actuators.

- If the normal force requirement changes during a MR finishing process that uses a magnetic tool based on a permanent magnet, the magnetic tool must be replaced with a permanent magnet of an appropriate strength. The electromagnet-based MRH technique is useful for a number of applications where MRAFF or R-MRAFF would fail, including the finishing of the interior surface of cylindrical workpiece made up of ferromagnetic materials.

REFERENCES

1. Rhoades, L.J., 1988. Abrasive flow machining[AFM]. *Manufacturing Engineering, 101*(5), pp.75–78.
2. Grover, V. and Singh, A.K., 2018. Improved magnetorheological honing process for nanofinishing of variable cylindrical internal surfaces. *Materials and Manufacturing Processes, 33*(11), pp.1177–1187.
3. Yamaguchi, H., Srivastava, A.K., Tan, M. and Hashimoto, F., 2014. Magnetic Abrasive Finishing of cutting tools for high-speed machining of titanium alloys. *CIRP Journal of Manufacturing Science and Technology, 7*(4), pp.299–304.
4. Yamaguchi, H., Kang, J. and Hashimoto, F., 2011. Metastable austenitic stainless steel tool for magnetic abrasive finishing. *CIRP Annals, 60*(1), pp.339–342.
5. Sirwal, S.A., Singh, A.K. and Paswan, S.K., 2020. Experimental analysis of magnetorheological finishing of blind hole surfaces using permanent magnet designed tools. *Journal of the Brazilian Society of Mechanical Sciences and Engineering, 42*(3), pp.1–23.
6. Baron, Y.M., 1975. Technology of abrasive machining in a magnetic field. *Mashinostroenie, Leningrad*, p.43.
7. Ruben, H.J., 1987. *Advances in Surface Treatment*, vol. 5. Niku-Lari, A., editor. Pergamon Press. pp. 239–256.
8. Kyoei Denko Co., Ltd. 2002. Ultra-precision internal finishing process for flexible pipes. *Mechanical Engineering*, 30(9), p.79. The Nikkan Kogyo Sinbun, Ltd.
9. Bedi, T.S. and Singh, A.K., 2016. Magnetorheological methods for nanofinishing–a review. *Particulate Science and Technology, 34*(4), pp.412–422.
10. Mali, H.S. and Manna, A., 2009. Current status and application of abrasive flow finishing processes: a review. *Proceedings of the Institution of Mechanical Engineers, Part B: Journal of Engineering Manufacture, 223*(7), pp.809–820.
11. Dixit, N., Sharma, V. and Kumar, P., 2021. Research trends in abrasive flow machining: a systematic review. *Journal of Manufacturing Processes, 64*, pp.1434–1461.
12. Jain, V. K., 2008. Advanced (non-traditional) machining processes. In *Machining*. Davim, J. P., editor. London: Springer-Verlag London Limited. pp. 299–325.
13. Rhoades, L., 1991. Abrasive flow machining: a case study. *Journal of Materials Processing Technology, 28*(1–2), pp.107–116.
14. Williams, R.E. and Rajurkar, K.P., 1992. Stochastic modeling and analysis of abrasive flow machining. *Journal of Engineering for Industry, 114*(1), pp.74–81.
15. Gorana, V.K., Jain, V.K. and Lal, G.K., 2004. Experimental investigation into cutting forces and active grain density during abrasive flow machining. *International Journal of Machine Tools and Manufacture, 44*(2–3), pp.201–211.
16. Walia, R.S., Shan, H.S. and Kumar, P., 2006. Abrasive flow machining with additional centrifugal force applied to the media. *Machining Science and Technology, 10*(3), pp.341–354.

17. Walia, R.S., Shan, H.S. and Kumar, P., 2006. Parametric optimization of centrifugal force-assisted abrasive flow machining (CFAAFM) by the Taguchi method. *Materials and Manufacturing Processes*, *21*(4), pp.375–382.
18. Reddy, M.K., Sharma, A.K. and Kumar, P., 2008. Some aspects of centrifugal force assisted abrasive flow machining of 2014 Al alloy. *Proceedings of the Institution of Mechanical Engineers, Part B: Journal of Engineering Manufacture*, *222*(7), pp.773–783.
19. Sankar, M.R., Jain, V.K. and Ramkumar, J., 2009. Experimental investigations into rotating workpiece abrasive flow finishing. *Wear*, *267*(1–4), pp.43–51.
20. Sankar, M.R., Jain, V.K. and Ramkumar, J., 2016. Nano-finishing of cylindrical hard steel tubes using rotational abrasive flow finishing (R-AFF) process. *The International Journal of Advanced Manufacturing Technology*, *85*(9), pp.2179–2187.
21. Sankar, M.R., Mondal, S., Ramkumar, J. and Jain, V.K., 2009. Experimental investigations and modeling of drill bit-guided abrasive flow finishing (DBG-AFF) process. *The International Journal of Advanced Manufacturing Technology*, *42*(7–8), pp.678–688.
22. Sankar, M.R., Mondal, S., Ramkumar, J. and Jain, V.K., 2009. Experimental investigations and modeling of drill bit-guided abrasive flow finishing (DBG-AFF) process. *The International Journal of Advanced Manufacturing Technology*, *42*, pp.678–688.
23. Brar, B.S., Walia, R.S., Singh, V.P. and Sharma, M., 2013. A robust helical abrasive flow machining (HLX-AFM) process. *Journal of The Institution of Engineers (India): Series C*, *94*(1), pp.21–29.
24. Jones, A.R. and Hull, J.B., 1998. Ultrasonic flow polishing. *Ultrasonics*, *36*(1–5), pp.97–101.
25. Sharma, A.K., Venkatesh, G., Rajesha, S. and Kumar, P., 2015. Experimental investigations into ultrasonic-assisted abrasive flow machining (UAAFM) process. *The International Journal of Advanced Manufacturing Technology*, *80*(1), pp.477–493.
26. Dabrowski, L., Marciniak, M. and Szewczyk, T., 2006. Analysis of abrasive flow machining with an electrochemical process aid. *Proceedings of the Institution of Mechanical Engineers, Part B: Journal of Engineering Manufacture*, *220*(3), pp.397–403.
27. Brar, B.S., Walia, R.S. and Singh, V.P., 2015. Electrochemical-aided abrasive flow machining (ECA 2 FM) process: a hybrid machining process. *The International Journal of Advanced Manufacturing Technology*, *79*(1–4), pp.329–342.
28. Singh, S. and Shan, H.S., 2002. Development of magneto abrasive flow machining process. *International Journal of Machine Tools and Manufacture*, *42*(8), pp.953–959.
29. Thomas, N., Sikandar, A.I. and Bukkapatnam, S., 2016. Nanofinishing of biomedical implants. In *Nanofinishing Science and Technology, Basic and Advanced Finishing and Polishing Processes,* 1st ed., Jain, V.K. editor. pp. 569–616. CRC Press.
30. Guo, J., Au, K.H., Sun, C.N., Goh, M.H., Kum, C.W., Liu, K., Wei, J., Suzuki, H. and Kang, R., 2019. Novel rotating-vibrating magnetic abrasive polishing method for double-layered internal surface finishing. *Journal of Materials Processing Technology*, *264*, pp.422–437.
31. Dhull, S., Walia, R.S., Murtaza, Q. and Singh, M., 2019. Electrochemo-magneto abrasive flow machine setup fabrication and experimental investigation of the process along with mathematical modeling and optimization. *Independent Journal of Management & Production*, *10*(8), pp.1834–1849.
32. Kang, J. and Yamaguchi, H., 2012. Internal finishing of capillary tubes by magnetic abrasive finishing using a multiple pole-tip system. *Precision Engineering*, *36*(3), pp.510–516.
33. Jain, V.K., 2009. Magnetic field assisted abrasive based micro-/nano-finishing. *Journal of Materials Processing Technology*, *209*(20), pp.6022–6038.

34. Mori, T., Hirota, K. and Kawashima, Y., 2003. Clarification of magnetic abrasive finishing mechanism. *Journal of Materials Processing Technology*, *143*, pp.682–686.
35. Fox, M., Agrawal, K., Shinmura, T. and Komanduri, R., 1994. Magnetic abrasive finishing of rollers. *CIRP Annals*, *43*(1), pp.181–184.
36. Shinmura, T., Takazawa, K. and Hatano, E., 1985. Study on magnetic-abrasive process. Application to plane finishing. *Bulletin of the Japan Society of Precision Engineering*, *19*(4), pp.289–291.
37. Yamaguchi, H. and Shinmura, T., 2004. Internal finishing process for alumina ceramic components by a magnetic field assisted finishing process. *Precision Engineering*, *28*(2), pp.135–142.
38. Singh, D.K., Jain, V.K., Raghuram, V. and Komanduri, R., 2005. Analysis of surface texture generated by a flexible magnetic abrasive brush. *Wear*, *259*(7–12), pp.1254–1261.
39. Iqbal, F., & Jha, S. (2019). Experimental investigations into transient roughness reduction in ball-end magneto-rheological finishing process. *Materials and Manufacturing Processes*, *34*(2), pp.224–231.
40. Kang, J., George, A. and Yamaguchi, H., 2012. High-speed internal finishing of capillary tubes by magnetic abrasive finishing. *Procedia CIRP*, *1*, pp.414–418.
41. Laroux, K.G., 2008. Using magnetic abrasive finishing for deburring produces parts that perform well and look great. *Alluring and Deburring*, *60*. https://www.yumpu.com/en/document/read/11479441 Accessed 5 January 2023
42. Heng, L., Kim, Y.J. and Mun, S.D., 2017. Review of superfinishing by the magnetic abrasive finishing process. *High Speed Mach*, *3*(1), pp.42–55.
43. Yamaguchi, H., Shinmura, T. and Kaneko, T., 2000. Development of a new internal finishing process applying magnetic abrasive finishing by use of pole rotation system. *International Journal of the Japan Society for Precision Engineering*, *30*(4), pp.317–322.
44. Misra, A., Pandey, P.M. and Dixit, U.S., 2017. Modeling of material removal in ultrasonic assisted magnetic abrasive finishing process. *International Journal of Mechanical Sciences*, *131*, pp.853–867.
45. Yun, H., Han, B., Chen, Y and Liao, M., 2016. Internal finishing process of alumina ceramic tubes by ultrasonic-assisted magnetic abrasive finishing. *International Journal of Advance Manufacturing Technology (2016)*, *85*, pp.727–734.
46. Kim, J.D., Kang, Y.H., Bae, Y.H. and Lee, S.W., 1997. Development of a magnetic abrasive jet machining system for precision internal polishing of circular tubes. *Journal of Materials Processing Technology*, *71*(3), pp.384–393.
47 Jain, V.K., 2009. *Advanced Machining Processes*. Allied Publishers.
48. Girma, B., Joshi, S.S., Raghuram, M.V.G.S. and Balasubramaniam, R., 2006. An experimental analysis of magnetic abrasives finishing of plane surfaces. *Machining science and Technology*, *10*(3), pp.323–340.
49. Singh, G., Kumar, H., Kansal, H.K. and Srivastava, A., 2020. Effects of chemically assisted magnetic abrasive finishing process parameters on material removal of inconel 625 tubes. *Procedia Manufacturing*, *48*, pp.466–473.
50. Shinmura, T. and Yamaguchi, H., 1995. Study on a new internal finishing process by the application of magnetic abrasive machining: internal finishing of stainless steel tube and clean gas bomb. *JSME International Journal Series C, Dynamics, Control, Robotics, Design and Manufacturing*, *38*(4), pp.798–804.
51. Yamaguchi, H. and Shinmura, T., 1999. Study of the surface modification resulting from an internal magnetic abrasive finishing process. *Wear*, *225*, pp.246–255.
52 Shinmura, T., Yamaguchi, H. and Aizawa, T., 1993. A new internal finishing process of non-ferromagnetic tubing by the application of a magnetic field: the development of a unit finishing apparatus using permanent magnets. *International Journal of the Japan Society for Precision Engineering*, *27*(2), pp.132–137.

53 Deepak, B., Walia, R. and Suri, N., 2012. Effect of rotational motion on the flat work piece magnetic abrasive finishing. *International Journal of Surface Engineering and Materials Technology*, *2*, pp.50–54.
54. Sihag, N., Kala, P. and Pandey, P.M., 2015. Chemo assisted magnetic abrasive finishing: experimental investigations. *Procedia CIRP*, *26*, pp.539–543.
55. Shinmura, T., Hatano, E. and Takazawa, K., 1986. The development of magnetic-abrasive finishing and its equipment by applying a rotating magnetic field. *Bulletin of JSME*, *29*(258), pp.4437–4443.
56. Saraswathamma, K., 2014. Magnetorheological finishing: a review. *International Journal of Current Engineering and Technology, Special Issue-2.*
57. Jain, V.K. ed., 2016. *Nanofinishing Science and Technology: Basic and Advanced Finishing and Polishing Processes*. CRC Press.
58. Kordonski, W.I. and Jacobs, S.D., 1996. Magnetorheological finishing. *International Journal of Modern Physics B*, *10*(23n24), pp.2837–2848.
59. Khan, D.A., Alam, Z. and Iqbal, F., 2021. *Magnetic Field Assisted Finishing: Methods, Applications and Process Automation*. CRC Press.
60. Bedi, T.S. and Singh, A.K., 2018. A new magnetorheological finishing process for ferromagnetic cylindrical honed surfaces. *Materials and Manufacturing Processes*, *33*(11), pp.1141–1149.
61. Bedi, T.S. and Singh, A.K., 2018. Development of magnetorheological fluid-based process for finishing of ferromagnetic cylindrical workpiece. *Machining Science and Technology*, *22*(1), pp.120–146.
62. Grover, V., and Singh, A.K., 2018. Improved magnetorheological honing process for nanofinishing of variable cylindrical internal surfaces. *Materials and Manufacturing Processes*, *33*(11), pp.1177–1187.
63. Jha, S. and Jain, V.K., 2004. Design and development of the magnetorheological abrasive flow finishing (MRAFF) process. *International Journal of Machine Tools and Manufacture*, *44*(10), pp.1019–1029.
64. Jha, S. and Jain, V.K., 2006. Modeling and simulation of surface roughness in magnetorheological abrasive flow finishing (MRAFF) process. *Wear*, *261*(7–8), pp.856–866.
65. Paswan, S.K., Bedi, T.S. and Singh, A.K., 2017. Modeling and simulation of surface roughness in magnetorheological fluid based honing process. *Wear*, *376*, pp.1207–1221
66. Das, M., Jain, V.K. and Ghoshdastidar, P.S., 2012. Nanofinishing of flat workpieces using rotational–magnetorheological abrasive flow finishing (R-MRAFF) process. *The International Journal of Advanced Manufacturing Technology*, *62*(1), pp.405–420.
67. Das, M., Jain, V.K. and Ghoshdastidar, P.S., 2011. The out-of-roundness of the internal surfaces of stainless steel tubes finished by the rotational–magnetorheological abrasive flow finishing process. *Materials and Manufacturing Processes*, *26*(8), pp.1073–1084.
68. Paswan, S.K. and Singh, A.K., 2019. Analysis of surface finishing mechanism in a newly developed rotational magnetorheological honing process for its productivity improvement. *Wear*, *426*, pp.68–82.

8 Post-processing methods for surface enhancement of additively manufactured parts

Atul Singh Rajput, Manas Das, and Sajan Kapil

8.1 INTRODUCTION

Additive manufacturing (*AM*) has been rapidly evolving in the last few decades and has become an important tool for manufacturing complex parts with high accuracy and precision [1]. This technology involves the layer-by-layer addition of materials, such as polymers, metals, ceramics, and composites, to create three-dimensional objects from digital models, as shown in Figure 8.1. However, the surfaces of additively manufactured parts often exhibit roughness, porosity, and other defects that can adversely affect their functional performance [2]. Post-processing techniques have been developed to overcome these limitations, reduce surface roughness, and improve functional performance. They can be broadly classified into chemical, electrochemical, mechanical, and thermal methods, depending on the nature of the surface modification [3]. These techniques can be used individually or in combination to achieve the desired surface finish and functional properties. Chemical post-processing techniques involve using chemical solutions to dissolve or etch the surface layer of the part to remove roughness, burrs, and other defects. Electrochemical methods use an electrolyte solution and an electrical current to remove material from the surface of the part, resulting in a smoother finish. Furthermore, mechanical post-processing techniques include grinding, polishing, and sandblasting, which use mechanical force to remove surface defects and create a smoother finish. However, thermal post-processing techniques involve using heat to modify the surface properties of the part, such as through laser or electron beam irradiation. Each post-processing technique has its advantages and limitations. Chemical methods are easy to use and effective in removing surface defects, but they may result in a loss of material and changes in surface properties [4]. Electrochemical methods offer precise control over the surface finish and can be used to selectively remove material, but they require specialized equipment and expertise [5]. Mechanical methods are widely used and effective but can be time-consuming and require skilled labor [6]. Thermal methods offer unique advantages, such as the ability to modify the surface properties without affecting the

DOI: 10.1201/9781003496298-8

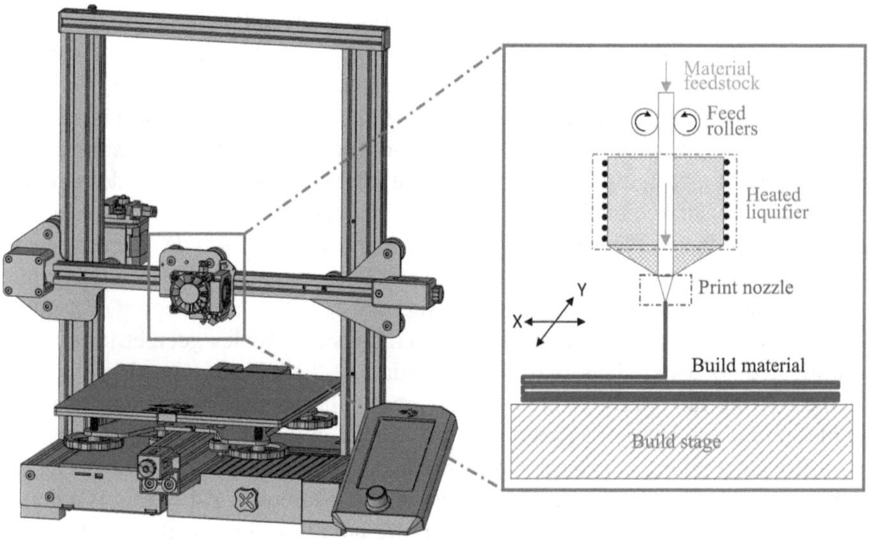

FIGURE 8.1 Fused deposition modeling.

bulk properties, but they require specialized equipment and expertise [7]. In recent years, there has been significant research on post-processing techniques for reducing surface roughness and improving the functional performance of 3D printed parts. This chapter will provide an overview of these techniques, their principles, materials, techniques, advantages, and limitations. Furthermore, it will survey the recent literature on these techniques and highlight their effectiveness in enhancing the surface quality and functional performance of additively manufactured parts.

8.2 DEFECTS IN ADDITIVELY MANUFACTURED PARTS

AM is a process that produces parts by layering materials to create a three-dimensional object. Despite its benefits, such as reduced material waste, lower manufacturing costs, and the ability to create complex geometries, AM is not immune to surface defects that can compromise part integrity and performance [8]. Different defects in additively manufacture, like warping, delamination, poor surface roughness, pillowing, Z-banding, and stringing, lead to altering the dimension accuracy of the fabricated part. Warping is the deformation of a part's shape due to thermal stresses during the build process. It typically occurs when there is a temperature difference between the part and the build platform, causing differential cooling and contraction. This defect is more common in materials with high thermal expansion coefficients, such as Acrylonitrile Butadiene Styrene (ABS) or polycarbonate, and can be mitigated by using a heated built platform, designing parts with thicker walls, and reducing the printing speed [9]. Moreover, delamination is a separation of the layers in a printed part, resulting in a weak bond between the layers. It is usually caused by insufficient adhesion between the layers due to a lack of heat, speed, or pressure. This defect is more common in materials such as Polylactic Acid PLA) or Polyethylene Terephthalate Glycol (PETG); however, it can be eliminated by increasing the bed temperature, printing slower, or using an adhesive or a raft [10]. Poor surface roughness is a common defect in AM parts and can occur due to several factors, such as layer height, printing speed, and nozzle size. It can affect the aesthetics and functionality of a part, especially if it has tight tolerances or needs to slide or rotate. To improve the surface quality of the additively manufactured parts, increasing the layer height, reducing the printing speed, and using a smaller nozzle size are beneficial. Furthermore, overhanging is another type of defect that occurs when the printer extrudes material in mid-air without any support, resulting in drooping or sagging [11]. This defect is common in parts with overhangs or complex geometries and can be mitigated by using support structures, reducing the printing speed, or increasing the cooling fan speed. Similarly, pillowing is the formation of small bulges or bubbles on the top layer of a part due to air trapped in the material [12]. It is caused by insufficient cooling or extrusion pressure and can be eliminated by increasing the cooling fan speed or decreasing the extrusion rate. Moreover, Z-banding is a defect that occurs when the layers of a printed part are not aligned correctly, resulting in a visible step-like pattern on the surface. This defect is more common in printers with non-rigid frames or poor mechanical components and can be mitigated by upgrading the printer's mechanical components or using improving the slicing quality of the computer-aided design (CAD) model [13]. Furthermore, stringing is a defect that occurs when the printer extrudes too much material between two points, resulting in thin threads or strings of material. This defect is more common in materials such as PLA or ABS and can be removed by reducing the printing temperature or increasing the retraction speed [14]. Surface defects are a common issue in AM parts and can affect the part's integrity and performance. Understanding the causes and mitigation strategies for each defect can improve the quality of AM parts and achieve better print results [15]. Hence new methods are required to reduce the surface roughness of the 3D printed parts followed by enhanced mechanical properties.

8.3 POST-PROCESSING METHOD

The present work discusses the post-processing methods available for surface enhancement of AM parts, their advantages and disadvantages, and their potential impact on the overall quality of the parts. The different surface post-processing methods are categorized based on the energy utilized during their operation, i.e., chemical, mechanical, and thermal, and are listed in Figure 8.2.

8.3.1 Chemical Polishing

In chemical polishing, a chemical solution is used to remove a thin layer of material from the surface of the part, resulting in a smoother and more polished surface [16]. Chemical polishing involves the dissolution of material from the surface of a part in a chemical solution. The technique primarily depends on the selective etching of the surface, with the etch rate being faster in regions with defects or roughness, as shown in Figure 8.3. The chemical solution used for polishing is typically an acidic or alkaline solution, depending on the type of material being polished. The chemical reaction

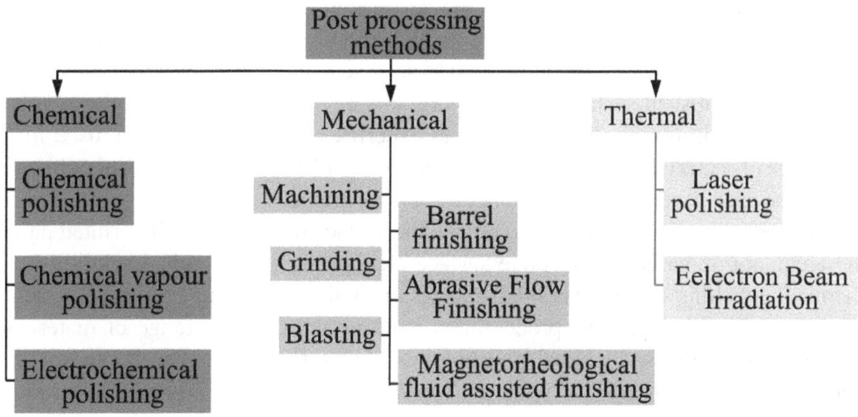

FIGURE 8.2 Different surface finishing processes.

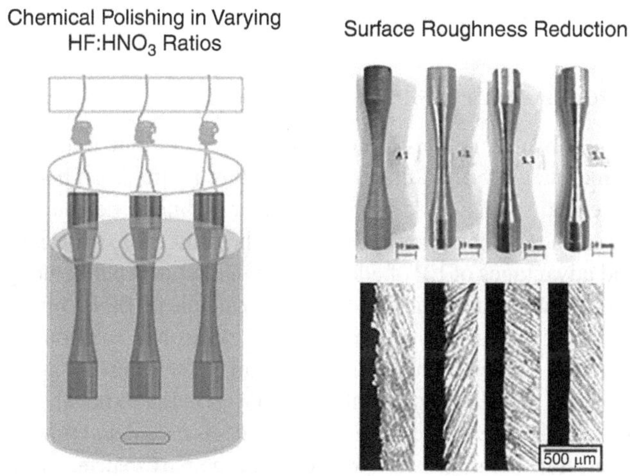

FIGURE 8.3 Chemical polishing of Ti-6Al-4V powder bed fusion produced part (reproduced with permission) [23].

between the solution and the material results in the formation of a thin layer of oxide or salt, which is then removed by rinsing the part with water [17]. The choice of chemical solution for polishing depends on the material of the part being polished. Metals, ceramics, and polymers all have different chemical properties and require different types of solutions for polishing. Some commonly used solutions for metal polishing include nitric acid, sulfuric acid, and phosphoric acid, while for polymer polishing, acetone or isopropyl alcohol may be used [18]. For ceramic parts, hydrofluoric acid or potassium hydroxide may be used [19]. There are several techniques for the chemical polishing of additively manufactured parts [20]. One commonly used technique is immersion polishing, where the part is immersed in the chemical solution for a certain period, followed by rinsing with water. Jet polishing is another method where the chemical solution is sprayed onto the surface of the part using a high-pressure jet, resulting in a more uniform polishing. Chemical polishing offers several advantages over other post-processing techniques. It can be used to achieve a high level of surface finish with a smooth and polished surface. Furthermore, it is a relatively simple and inexpensive process, requiring minimal equipment and expertise. Moreover, it can be used for a wide range of materials, including metals, ceramics, and polymers. Chemical polishing can also be used to remove surface defects such as scratches, cracks, and pores. Hence, chemical polishing is widely used to reduce the surface roughness of 3D printed parts. The technique is based on selective surface etching using a chemical solution [22]. Chemical polishing offers several advantages, including achieving good surface quality, a simple and inexpensive process, and working with a wide range of materials. Despite its advantages, chemical polishing also has some limitations. It can only be used to remove a thin layer of material, typically less than 10 microns. Moreover, it may not be suitable for parts with complex geometries or internal channels, as the solution may not be able to penetrate into these regions. Furthermore, the process can also result in the removal of material from regions with high curvature, resulting in a loss of dimensional accuracy [21]. However, the inability to remove large amounts of material is another limitation associated with chemical polishing.

8.3.2 Electrochemical polishing

Electrochemical polishing (EP) is a post-processing technique that can improve the surface finish and integrity of additively manufactured parts [5]. This technique uses an electric current to selectively dissolve surface roughness, defects, and impurities from the part's surface. This review will discuss the principles, materials, techniques, limitations, and advantages of EP of additively manufactured parts, along with a literature survey of recent research in this area. The principle of EP involves the selective anodic dissolution of the surface of the part in an electrolyte solution. The process is based on the concept of electrochemistry, where an electric current is passed through the part immersed in an electrolyte solution, resulting in the removal of material from the surface. The material removed during the process is calculated through the current density, the electrolyte solution's composition, and the part's surface morphology [5]. EP is suitable for a wide range of materials, including metals, alloys, and composites. Some commonly used materials for EP of additively manufactured parts include stainless steel, titanium alloys, and nickel-based alloys [24]. The choice of electrolyte solution depends on the material being polished, with some common solutions including sulfuric acid, phosphoric acid, and nitric acid. There are several techniques for EP of additively manufactured parts. One commonly used technique is electropolishing, where the part is immersed in an electrolyte solution, and an electric current is passed through the part, as shown in Figure 8.4 [25]. Another technique is pulse EP, where the current is pulsed, resulting in a more controlled and uniform polishing. Jet EP is another technique where the electrolyte solution is sprayed onto the surface of the part using a high-pressure jet, resulting in a more uniform polishing [4]. EP offers several advantages over other post-processing methods. It is a relatively simple and inexpensive process, requiring minimal equipment and expertise [26]. Furthermore, it can be used for a wide range of materials, including metals, alloys, and composites. EP can also be used to improve the part's corrosion resistance and mechanical

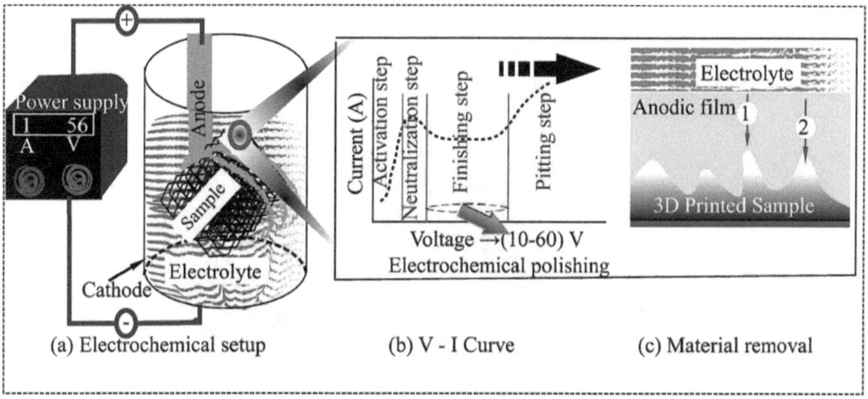

FIGURE.8.4 Working principle of electropolishing (reproduced with permission) [30].

properties. Despite its advantages, EP also has some limitations. It may not be suitable for parts with complex geometries or internal channels, as the electrolyte solution may not penetrate these regions [27]. Furthermore, the process can also remove material from regions with high curvature, resulting in a loss of dimensional accuracy [28]. Moreover, the process can also result in the formation of hydrogen gas, which can lead to hydrogen embrittlement and reduce the part's mechanical properties [29].

8.3.3 Computer Numerical Control (CNC) machining

In some features of AM, the layer height can be decreased for a better surface finish but to an extent. This will, in turn, increase the build time and is constrained to the available 3D printer. Surface roughness directly results from the chordal effect, staircase effect, inter-road width, and sub-perimeter voids. It worsens the surface quality of the fabricated parts. Inherent effects such as the staircase effect and sometimes warpage are inevitable. The staircase effect is the difference between CAD geometry and the final printed product. It occurs because the layer-by-layer slices approximate the geometry, and also due to the constraint of the extrusion nozzle [64]. To remove these, a number of methods are available depending upon the material deposited. CNC machining is the most straightforward technique to post-process the part. CNC machining is a post-processing technique that can reduce the surface irregularities of additively manufactured parts [6]. This technique involves the use of a computer-controlled cutting tool to remove material from the part's surface, resulting in a smoother and more uniform surface finish. The principle of CNC machining involves the use of a computer-controlled cutting tool to remove material from the part's surface, as shown in Figure 8.5 [31]. The process is based on the concept of subtractive manufacturing, where the material is detached from the stock of material to create the desired shape. In the case of additively manufactured parts, CNC machining can be used to reduce surface irregularities, defects, and excess material [32]. CNC machining is suitable for a wide range of materials, including metals, plastics, and composites. Some commonly used materials for CNC machining of additively manufactured parts include aluminum, titanium alloys, and carbon fiber-reinforced

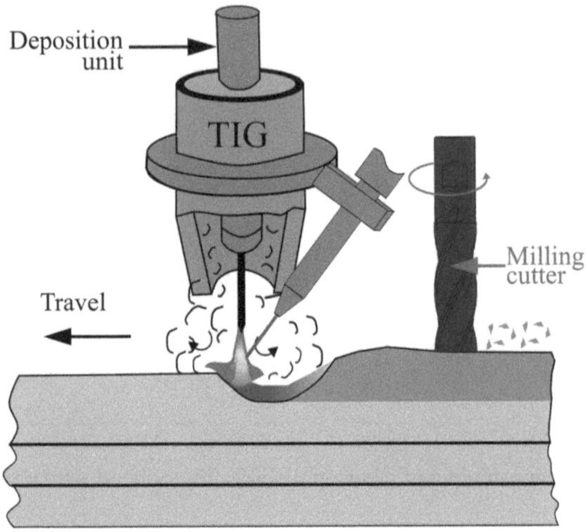

FIGURE 8.5 Wire arc additive manufactured part followed by CNC machining to improve the surface quality of the fabricated part.

polymers. The choice of cutting tool and machining parameters depends on the material being machined, with some common parameters, namely, depth of cut, feed rate, and cutting speed. Several techniques exist for the CNC machining of additively manufactured parts [33]. One commonly used technique is three-axis machining, where the cutting tool moves along three axes to detach the material from the part's surface. Another technique is five-axis machining, where the cutting tool can move along five axes (*X*, *Y*, *Z*, *A*, and *B*) to machine complex geometries and curved surfaces. Hybrid machining, combining additive and subtractive manufacturing, is also becoming increasingly popular [34]. CNC machining offers several advantages over other post-processing techniques. It also produces a good surface quality with uniformity. Moreover, it can also be used to remove excess material and improve the part's dimensional accuracy [35]. Furthermore, it is a relatively fast and efficient process, requiring minimal manual intervention. Finally, it can be used for a wide range of materials, including metals, plastics, and composites. Despite its advantages, CNC machining also has some limitations. It can be time-consuming and expensive, especially for parts with complex geometries. Moreover, it can result in the loss of material and reduce the mechanical properties of the part [36]. Furthermore, it may not be suitable for parts with internal channels or delicate features, as the cutting tool may not be able to access these regions. Recent research has focused on improving the effectiveness and efficiency of CNC machining for reducing the surface roughness value of 3D printed parts [37]. This technique produces uniform surface quality, removes excess material, and enhances the dimensional accuracy of the part.

8.3.4 Barrel Finishing

Barrel finishing is a commonly used mass-finishing technique that can enhance the surface quality of 3D printed parts [38]. This technique involves placing the parts to be finished in a rotating barrel filled with abrasive media, which removes surface roughness and improves the part's surface finish, as shown in Figure 8.6. The principle of barrel finishing involves placing the parts to be finished in a rotating barrel filled with abrasive media [39]. The abrasive media, which can be made of materials such as ceramic, plastic, or steel, rubs against the part's surface, removing surface roughness and improving the surface finish. The process is based on the concept of mass finishing, where a large number of parts can be finished at the same time. Barrel finishing is suitable for a wide range of materials, including metals, plastics, and composites [40]. Some commonly used materials for barrel finishing of additively manufactured parts include titanium alloys, stainless steel, and aluminum. The choice of abrasive media depends on the material being finished, with some common media including ceramic balls, plastic chips, and steel pins. There are several techniques for barrel finishing of additively manufactured parts [41]. One commonly used technique is vibratory finishing, where the barrel is vibrated to enhance the abrasive action of the media. Another technique is centrifugal barrel finishing, where the barrel is rotated at high speeds to create a centrifugal force that enhances the finishing action of the media. Magnetic barrel finishing, where the abrasive media is magnetized to improve the finishing action, is also becoming increasingly popular. Barrel finishing offers several advantages over other post-processing techniques [37]. It can be used to produce a highly polished surface. Moreover, it can be used to finish a huge number of workpieces simultaneously, making it a highly efficient process. Furthermore, it is a relatively very cost-efficient and simple process, requiring minimal manual intervention. Also, it can be used for a wide range of materials, including plastics, metals, and composites. Despite its advantages, barrel finishing also has some limitations. It may not be suitable for parts with delicate features or internal channels, as the abrasive media may not be able to access these regions. Furthermore, it may result in material loss and reduce the part's mechanical properties. Moreover, it may not be suitable for parts with complex geometries, as the abrasive media may not be able to reach all parts of the surface [42].

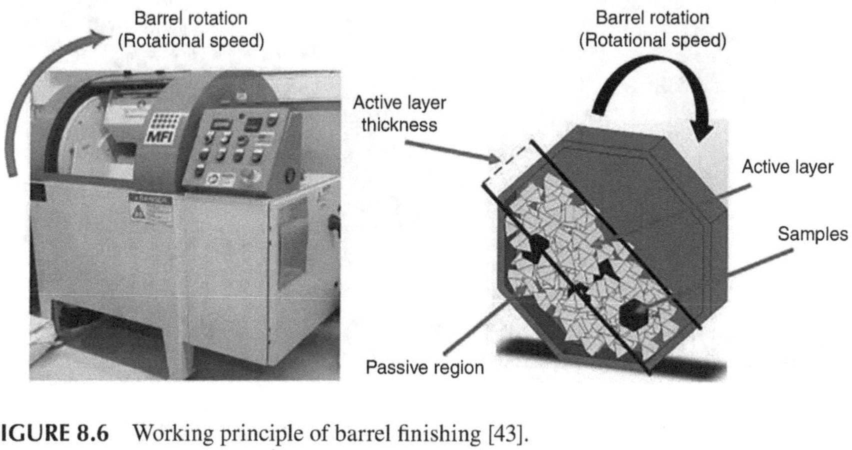

FIGURE 8.6 Working principle of barrel finishing [43].

8.3.5 Magnetorheological finishing

Magnetorheological finishing (MRF) is an advanced polishing technique that uses a magnetorheological fluid to remove surface irregularities and enhance the part's quality of 3D fabricated parts [44]. The principle of MRF involves using a magnetorheological fluid, which is a suspension of magnetic particles in a carrier fluid. When a magnetic field is applied to the fluid, the particles align along the field lines, increasing the viscosity and making the fluid more abrasive, as shown in Figure 8.7 [45]. The fluid is then applied to the part's surface to be finished, and a magnetic field is applied to the fluid, causing the abrasive particles to reduce the surface irregularities and improve the surface finish [46]. MRF is suitable for various materials, including metals, ceramics, and glasses [47]. Some commonly used materials for MRF of additively manufactured parts include stainless steel, titanium alloys, and aluminum [48]. The choice of magnetorheological fluid and magnetic field strength depends on the finished material, with some common fluids including iron oxide and carbonyl iron particles [49]. There are several techniques for MRF of additively manufactured parts. One commonly used technique is rotary MRF, where the part to be finished is rotated while the magnetorheological fluid is applied. Another technique is reciprocating MRF, where the part is moved back and forth while the fluid is applied [50]. Another technique is fixed abrasive MRF, where a fixed abrasive is added to the magnetorheological fluid to improve the finishing action. MRF offers several advantages over other post-processing techniques. It can also be used to produce uniform mirror-like polished surface [51]. Furthermore, it can be used to finish parts

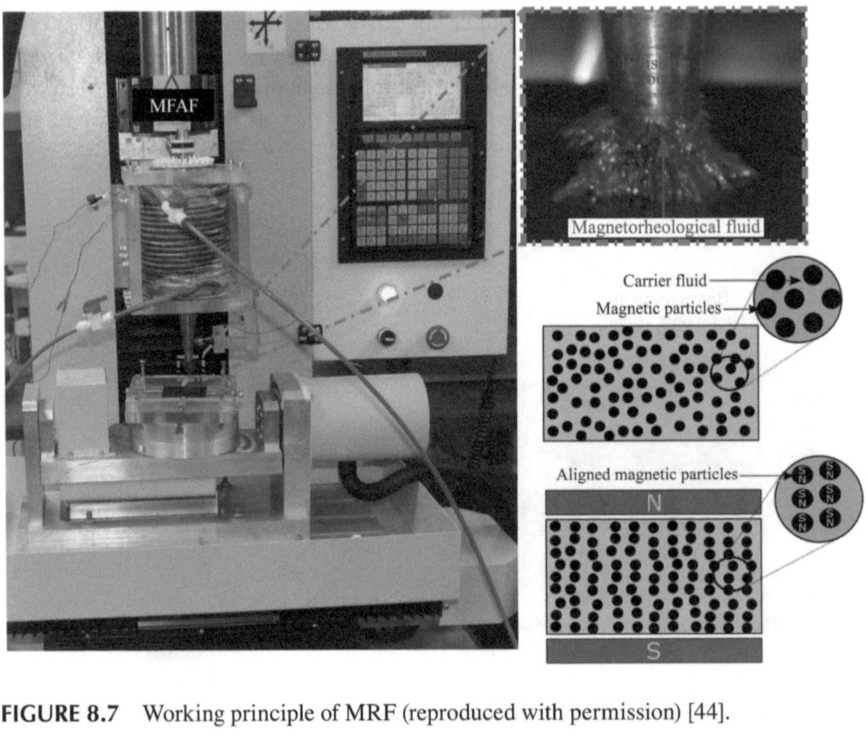

FIGURE 8.7 Working principle of MRF (reproduced with permission) [44].

with complicated shapes and geometries, as the magnetorheological fluid can access hard-to-reach areas [52]. Moreover, it can be used for various materials, including metals, ceramics, and glasses. Also, it is a relatively simple and low-cost process, requiring minimal manual intervention [53]. Despite its advantages, MRF also has some limitations. Moreover, it may not be suitable for parts with delicate features or internal channels, as the abrasive particles may be unable to access these regions. Furthermore, it may result in the loss of material and reduce the part's mechanical properties. Finally, the choice of magnetorheological fluid and magnetic field strength needs to be carefully optimized for each material, which can be costly and time-consuming [54].

8.3.6 Laser polishing

Laser polishing (LP) is a promising technique for reducing the surface irregularities of additively manufactured parts [7]. It is a process that uses a laser beam to melt and resolidify the surface of a part, resulting in a smooth and polished surface [55]. The laser beam is focused on the surface of the part, and the energy of the beam melts a thin layer of material, as shown in Figure 8.8 [56]. As the material cools and solidifies, it forms a smooth and uniform surface. LP is suitable for a wide range of materials, including metals, ceramics, and glasses [57]. Some commonly used materials for LP of additively manufactured parts include stainless steel, titanium alloys, and aluminum [58]. The choice of laser parameters, such as power and scanning speed, depends the material being polished. There are several techniques for LP additively manufactured parts. One commonly used technique is continuous wave LP, where the laser beam is continuously applied to the surface of the part. Another technique is pulsed LP, where the laser beam is pulsed to achieve a more precise and controlled melting and resolidification of the surface. LP offers several advantages over other post-processing techniques [59]. It can also be used to produce uniform surface quality. Furthermore, it can finish parts with complex geometries, as the laser beam can access hard-to-reach areas. Moreover, it can be used for various materials, including metals, ceramics, and glasses. Finally, it is a relatively simple and low-cost

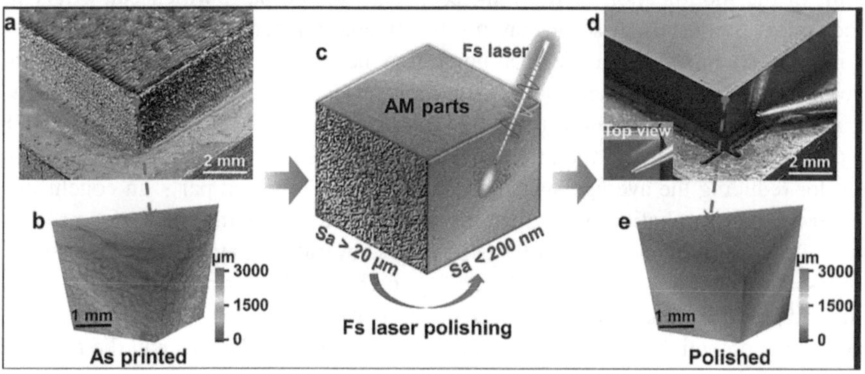

FIGURE 8.8 Laser polished surface (reproduced with permission) [63].

process, requiring minimal manual intervention [60]. Despite its advantages, LP also has some limitations. It may result in the loss of material and reduce the mechanical properties of the part [61]. Moreover, it may not be suitable for parts with delicate features or internal channels, as the laser beam may not be able to access these regions. Also, the choice of laser parameters needs to be carefully optimized for each material, which can be time-consuming and expensive [62]. Recent research has focused on improving the effectiveness and efficiency of LP to reduce the surface irregularities of the 3D fabricated parts [63].

8.3.7 Electron Beam Irradiation

Electron beam irradiation is the most efficient technique for enhancing the surface quality of additively manufactured parts [65]. Electron beam irradiation is a process that uses a beam of high-energy electrons to irradiate the surface of a part, resulting in the modification of the surface properties. The high-energy electrons interact with the atoms of the surface, causing structural changes and inducing new surface features. Electron beam irradiation can be used for various materials, including metals, ceramics, and polymers [66]. Some commonly used materials for electron beam irradiation of additively manufactured parts include titanium alloys, stainless steel, and aluminum. The choice of electron beam parameters, such as energy and dose, depends on the irradiated material. There are several techniques for electron beam irradiation of additively manufactured parts. One commonly used technique is continuous wave electron beam irradiation, where the electron beam is continuously applied to the surface of the part [67]. Another technique is pulsed electron beam irradiation, where the electron beam is pulsed to achieve a more precise and controlled surface modification. Electron beam irradiation offers several advantages over other post-processing techniques. It can be used to modify the surface properties of the part, including its hardness, wear resistance, and corrosion resistance. Moreover, it can be used for a smooth and polished surface. Furthermore, it can be used for various materials, including metals, ceramics, and polymers. Finally, it is a relatively simple and low-cost process, requiring minimal manual intervention. Despite its advantages, electron beam irradiation also has some limitations. It may result in the modification of the bulk properties of the part, which can affect its mechanical properties. Also, it may not be suitable for parts with delicate features or internal channels, as the electron beam may not be able to access these regions. Moreover, the choice of electron beam parameters needs to be carefully optimized for each material, which can be time-consuming and expensive [68]. Recent research has focused on improving the effectiveness and efficiency of electron beam irradiation for reducing the average surface roughness of 3D printed parts. In conclusion, electron beam irradiation is an efficient technique for reducing the surface roughness of additively manufactured parts. It offers several advantages over other post-processing techniques, including the ability to change the surface properties of the part and produce a high level of surface finish. However, it also has some limitations, such as the potential modification of the bulk properties of the part and the need for careful optimization of electron beam parameters.

8.4 CONCLUSIONS

The present work aims to analyze the influence of the various post-processing methods to improve the surface quality of the additively manufactured parts. The categorization is made based on the energy source used during the surface finishing operation, i.e., chemical, mechanical, and thermal energy. The major finding of the research work is summarized below:

- Chemical polishing is a widely used post-processing technique for improving the surface finish of additively manufactured parts. The technique is based on selective surface etching using a chemical solution. Chemical polishing offers several advantages, including the ability to achieve a high level of surface finish, a simple and inexpensive process, and the ability to work with a wide range of materials. However, the technique also has some limitations, including the inability to remove large amounts of material and the limitations.
- Barrel finishing is an extensively used mass-finishing technique for reducing surface irregularities of 3D printed parts. This technique can be used to achieve a high level of surface finish and finish a large number of parts simultaneously.
- EP is a widely used post-processing technique for improving the surface finish with minimum surface finishing time. However, an adverse effect of electrolytes on the polished surfaces are major challenge.
- CNC machining is a widely used post-processing technique for improving the surface quality of additively manufactured parts. This technique can be used to achieve a high level of surface finish, remove excess material, and improve the dimensional accuracy of the part.
- Electron beam irradiation is a promising technique for improving the surface quality of additively manufactured parts. It offers several advantages over other post-processing techniques, including the ability to enhance the surface properties of the part and produce a high level of surface finish. However, it also has some limitations, such as the potential modification of the bulk properties of the part and the need for careful optimization of electron beam parameters.
- MRF is capable of producing surface roughness value in the range of a few nanometers and produces a mirror-like polished surface without altering the surface topography of the workpiece. However, the slower polishing rate is a major limitation of the process.

ACKNOWLEDGMENT

We acknowledge Science & Engineering Research Board (SERB), New Delhi, India, for their financial support for project No. EEQ/2017/000597 entitled “Fabrication of Prosthetic Implants and further Nanofinishing Using Magnetic Field Assisted Finishing (MFAF) Process.”

REFERENCES

1. Y. Fang, S. Attarilar, Z. Yang, G. Wei, Y. Fu, and L. Wang, "Toward bactericidal enhancement of additively manufactured titanium implants," *Coatings*, vol. 11, no. 6, p. 668, May 2021, doi: 10.3390/COATINGS11060668.
2. T. Thenard, A. Catapano, R. Allena, M. El May, N. Saintier, and M. Mesnard, "Topography and wettability characterization of surfaces manufactured by SLM and treated by chemical etching," *Mech. Adv. Mater. Struct.*, vol. 29, no. 12, pp. 1674–1691, 2020, doi: 10.1080/15376494.2020.1836292.
3. K. L. Tan and S. H. Yeo, "Surface modification of additive manufactured components by ultrasonic cavitation abrasive finishing," *Wear*, vol. 378–379, pp. 90–95, May 2017, doi: 10.1016/J.WEAR.2017.02.030.
4. V. S. Muthaiah, S. Indrakumar, S. Suwas, and K. Chatterjee, "Surface engineering of additively manufactured titanium alloys for enhanced clinical performance of biomedical implants: A review of recent developments," *Bioprinting*, vol. 25, p. e00180, 2022.
5. J. Luo *et al.*, "Electropolishing influence on biocompatibility of additively manufactured Ti-Nb-Ta-Zr: In vivo and in vitro," *J. Mater. Sci.: Mater. Med.*, vol. 34, no. 5, p. 25, 2023.
6. L. An, D. Wang, and D. Zhu, "Combined electrochemical and mechanical polishing of interior channels in parts made by additive manufacturing," *Addit. Manuf.*, vol. 51, p. 102638, Mar. 2022, doi: 10.1016/J.ADDMA.2022.102638.
7. C. P. Ma, Y. C. Guan, and W. Zhou, "Laser polishing of additive manufactured Ti alloys," *Opt. Lasers Eng.*, vol. 93, pp. 171–177, Jun. 2017, doi: 10.1016/J.OPTLASENG.2017.02.005.
8. N. Sanaei and A. Fatemi, "Defects in additive manufactured metals and their effect on fatigue performance: A state-of-the-art review," *Prog. Mater. Sci.*, vol. 117, p. 100724, 2021.
9. Q. Lu and C. H. Wong, "Applications of non-destructive testing techniques for post-process control of additively manufactured parts," *Virtual Phys. Prototyp.*, vol. 12, no. 4, pp. 301–321, Oct. 2017, doi: 10.1080/17452759.2017.1357319.
10. A. Du Plessis, I. Yadroitsava, and I. Yadroitsev, "Effects of defects on mechanical properties in metal additive manufacturing: A review focusing on X-ray tomography insights," *Mater. Des.*, vol. 187, p. 108385, 2020.
11. Z. Snow, E. Reutzel, and J. Petrich, "Correlating in-situ sensor data to defect locations and part quality for additively manufactured parts using machine learning," *J. Mater. Process. Technol.*, vol. 302, p. 117476, 2022.
12. X. Peng *et al.*, "The potency of defects on fatigue of additively manufactured metals," *Int. J. Mech. Sci.*, vol. 221, p. 107185, 2022.
13. A. du Plessis and S. G. le Roux, "Standardized X-ray tomography testing of additively manufactured parts: A round robin test," *Addit. Manuf.*, vol. 24, pp. 125–136, 2018.
14. E. Wycisk, A. Solbach, S. Siddique, D. Herzog, F. Walther, and C. Emmelmann, "Effects of defects in laser additive manufactured Ti-6Al-4V on fatigue properties," *Phys. Procedia*, vol. 56, pp. 371–378, 2014.
15. B. Wu *et al.*, "A review of the wire arc additive manufacturing of metals: properties, defects and quality improvement," *J. Manuf. Process.*, vol. 35, pp. 127–139, Oct. 2018, doi: 10.1016/J.JMAPRO.2018.08.001.
16. B. Van Hooreweder, K. Lietaert, B. Neirinck, N. Lippiatt, and M. Wevers, "CoCr F75 scaffolds produced by additive manufacturing: Influence of chemical etching on powder removal and mechanical performance," *J. Mech. Behav. Biomed. Mater.*, vol. 70, pp. 60–67, Jun. 2017, doi: 10.1016/J.JMBBM.2017.03.017.
17. J. C. Kim and I. S. L. Yeo, "Bone response to conventional titanium implants and new zirconia implants produced by additive manufacturing," *Materials*, vol. 14, no. 16, p. 4405, Aug. 2021, doi: 10.3390/MA14164405.

18. M. Y. Mommaerts, "Evolutionary steps in the design and biofunctionalization of the additively manufactured sub-periosteal jaw implant 'AMSJI' for the maxilla," *Int. J. Oral Maxillofac. Surg.*, vol. 48, no. 1, pp. 108–114, Jan. 2019, doi: 10.1016/J.IJOM.2018.08.001.
19. A. N. Aufa, M. Z. Hassan, and Z. Ismail, "Recent advances in Ti-6Al-4V additively manufactured by selective laser melting for biomedical implants: Prospect development," *J. Alloys Compd.*, vol. 896, p. 163072, Mar. 2022, doi: 10.1016/J.JALLCOM.2021.163072.
20. Y. Y. Sun *et al.*, "Fatigue performance of additively manufactured Ti-6Al-4V: Surface condition vs. internal defects," *JOM*, vol. 72, no. 3, pp. 1022–1030, Mar. 2020, doi: 10.1007/S11837-020-04025-7.
21. A. Kopp *et al.*, "Influence of design and postprocessing parameters on the degradation behavior and mechanical properties of additively manufactured magnesium scaffolds," *Acta Biomater.*, vol. 98, pp. 23–35, Oct. 2019, doi: 10.1016/J.ACTBIO.2019.04.012.
22. M. Lowther *et al.*, "Clinical, industrial, and research perspectives on powder bed fusion additively manufactured metal implants," *Addit. Manuf.*, vol. 28, pp. 565–584, Aug. 2019, doi: 10.1016/J.ADDMA.2019.05.033.
23. M. Bezuidenhout, G. Ter Haar, T. Becker, S. Rudolph, O. Damm, and N. Sacks, "The effect of HF-HNO3 chemical polishing on the surface roughness and fatigue life of laser powder bed fusion produced Ti6Al4V," *Mater. Today Commun.*, vol. 25, p. 101396, Dec. 2020, doi: 10.1016/J.MTCOMM.2020.101396.
24. Y. Zhang, J. Li, S. Che, and Y. Tian, "Electrochemical polishing of additively manufactured Ti–6Al–4V alloy," *Met. Mater. Int.*, vol. 26, pp. 783–792, 2020.
25. A. Bernhardt, J. Schneider, A. Schroeder, K. Papadopoulous, E. Lopez, F. Brückner, and U. Botzenhart, "Surface conditioning of additively manufactured titanium implants and its influence on materials properties and in vitro biocompatibility," *Mater. Sci. Eng. C.*, vol. 119, p. 111631, 2021.
26. Y. C. Wu, C. N. Kuo, Y. C. Chung, C. H. Ng, and J. C. Huang, "Effects of electropolishing on mechanical properties and bio-corrosion of Ti6Al4V fabricated by electron beam melting additive manufacturing," *Materials*, vol. 12, no. 9, p. 1466, 2019, doi: 10.3390/ma12091466.
27. Z. Chaghazardi, R. Wüthrich, "Electropolishing of additive manufactured metal parts," *J. Electrochem. Soc.*, vol. 169, no. 4, p. 043510, 2022.
28. M. M. Rahman, L. A. Hof, B. Johnston, S. A. Khanoki, D. Pasini, and R. Wüthrich, "Electropolishing of additive manufactured porous titanium for medical implants," In *Electrochemical Society Meeting Abstracts 227*, no. 17, pp. 1247–1247, 2015, The Electrochemical Society, Inc.
29. N. Vanderesse, P. Bocher, N. Nuño, A. Yánez, and L. A. Hof, "On the characterization of roughness and geometrical irregularities of additively manufactured single titanium-alloy struts," *Addit. Manuf.*, vol. 54, p. 102731, 2022.
30. M. M. Basha, S. M. Basha, V. K. Jain, and M. R. Sankar, "State of the art on chemical and electrochemical based finishing processes for additive manufactured features," *Addit. Manuf.*, vol. 58, p. 103028, Oct. 2022, doi: 10.1016/J.ADDMA.2022.103028.
31. S. Kapil, A. S. Rajput, and R. Sarma, "Hybridization in wire arc additive manufacturing," *Front. Mech. Eng.*, vol. 8, p. 96, Sep. 2022, doi: 10.3389/FMECH.2022.981846/BIBTEX.
32. M. Deja, D. Zielí Nski, A. Zuhra, A. Kadir, and S. N. Humaira, "Applications of additively manufactured tools in abrasive machining—A literature review," *Materials*, vol. 14, no. 5, p. 1318, 2021, doi: 10.3390/ma14051318.
33. W. Tillmann, L. Hagen, D. Stangier, N. F. L. Dias, J. Görtz, and M. D. Kensy, "Lapping and polishing of additively manufactured 316L substrates and their effects on the microstructural evolution and adhesion of PVD CrAlN coatings," *Surf. Coat. Technol.*, vol. 428, p. 127905, 2021.

34. F. F. Fashanu, D. J. Marcellin-Little, and B. S. Linke, "Review of surface finishing of additively manufactured metal implants," In *International Manufacturing Science and Engineering Conference*, vol. 84256, p. V001T03A013, 2020, American Society of Mechanical Engineers.
35. Q. Ai, J. Khosravi, B. Azarhoushang, A. Daneshi, and B. Becker, "Digital light processing-based additive manufacturing of resin bonded SiC grinding wheels and their grinding performance," *Int. J. Adv. Manuf. Technol.*, vol. 118, no. 5–6, pp. 1641–1657, Jan. 2022, doi: 10.1007/S00170-021-08016-Z.
36. B. Denkena *et al.*, "Additive manufacturing of metal-bonded grinding tools," *Int. J. Adv. Manuf. Technol.*, vol. 107, no. 5–6, pp. 2387–2395, Mar. 2020, doi: 10.1007/S00170-020-05199-9.
37. A. Khan, *et al.*, "Development of barrel finishing machine to improve surface finish of the wire arc additive manufactured parts," *Procedia CIRP*, vol. 91, pp. 330–335, 2020.
38. J. Lee, A. Nagalingam, and S. H. Yeo, "A review on the state-of-the-art of surface finishing processes and related ISO/ASTM standards for metal additive manufactured components," *Virtual Phys. Prototyp.*, vol. 16, no. (1), pp. 68–96, 2021.
39. A. Boschetto, L. Bottini, and F. Veniali, "Surface roughness and radiusing of Ti6Al4V selective laser melting-manufactured parts conditioned by barrel finishing," *Int. J. Adv. Manuf. Technol.*, vol. 94, no. 5–8, pp. 2773–2790, Feb. 2018, doi: 10.1007/s00170-017-1059-6.
40. F. Tamburrino, S. Barone, A. Paoli, and A. V. Razionale, "Post-processing treatments to enhance additively manufactured polymeric parts: A review," *Virtual Phys. Prototyp.*, vol. 16, no. (2), pp. 221–254, 2021.
41. A. W. Hashmi, H. S. Mali, A. Meena, A. P. V. Puerta, and M. E. Kunkel, "Surface characteristics improvement methods for metal additively manufactured parts: A review," *Adv. Mater. Process. Technol.*, vol. 8, no. 4, pp. 4524–4563, 2022.
42. M. Jamal and M. N. Morgan, "Design process control for improved surface finish of metal additive manufactured parts of complex build geometry," *Inventions*, vol. 2, no. 4, p. 36, 2017.
43. M. Khorasani, I. Gibson, A. H. Ghasemi, M. Brandt, and M. Leary, "On the role of wet abrasive centrifugal barrel finishing on surface enhancement and material removal rate of LPBF stainless steel 316L," *J. Manuf. Process.*, vol. 59, pp. 523–534, Nov. 2020, doi: 10.1016/J.JMAPRO.2020.09.058.
44. A. S. Rajput, M. Das, and S. Kapil, "Investigation of surface characteristics on post processed additively manufactured biomaterial through magnetorheological fluid assisted finishing process," *Wear*, vol. 522, p. 204684, Jun. 2023, doi: 10.1016/J.WEAR.2023.204684.
45. M. Al-Amin *et al.*, "Investigation of coatings, corrosion and wear characteristics of machined biomaterials through hydroxyapatite mixed-EDM process: A review," *Materials*, vol. 14, no. 13, p. 3597, Jun. 2021, doi: 10.3390/MA14133597.
46. A. S. Rajput, S. Kapil, and M. Das, "Methods for surface superfinishing of prosthesis," In Shrikrishna Nandkishor Joshi and Pranjal Chandra (eds.), *Advanced Micro- and Nano-manufacturing Technologies*, Singapore: Springer, pp. 335–360, 2022, doi: 10.1007/978–981-16–3645-5_15.
47. A. S. Rajput, A. Singh, S. Kapil, and M. Das, "Investigations on the toolpath strategies for CNC magnetorheological fluid assisted finishing (MFAF) process," *Int. J. Adv. Manuf. Technol.*, vol. 121, no. 1, pp. 949–966, May 2022, doi: 10.1007/S00170-022-09307-9/TABLES/4.
48. A. S. Rajput, S. Kapil, and M. Das, "Computer-aided process planning system for super finishing of flat surfaces with pockets through magnetorheological finishing process," *Int. J. Comput. Integr. Manuf.*, pp. 1–17, Mar. 2023, doi: 10.1080/0951192X.2023.2189313.

49. A. S. Rajput, S. Kapil, and M. Das, "Computational techniques for predicting process parameters in the magnetorheological fluid-assisted finishing Process," In A. Kumar, Y. Gori, N. Dutt, Y. K. Singla, and A. Maurya (Eds.), *Advanced Computational Methods in Mechanical and Materials Engineering*, Boca Raton: CRC Press, pp. 125–146, Oct. 2021, doi: 10.1201/9781003202233-10.
50. M. Das, V. K. Jain, and P. S. Ghoshdastidar, "A 2D CFD simulation of MR polishing medium in magnetic field-assisted finishing process using electromagnet," *Int. J. Adv. Manuf. Technol.*, vol. 76, no. 1–4, pp. 173–187, 2014, doi: 10.1007/s00170-014-5847-y.
51. A. S. Rajput, M. Das, and S. Kapil, "A comprehensive review of magnetorheological fluid assisted finishing processes," *Mach. Sci. Technol.*, vol. 26, no. 3, pp. 339–376, 2022, doi: 10.1080/10910344.2022.2129982.
52. A. S. Rajput, M. Das, and S. Kapil, "Investigations on a hybrid chemo-magnetorheological finishing process for freeform surface quality enhancement," *J. Manuf. Process.*, vol. 81, pp. 522–536, Sep. 2022, doi: 10.1016/J.JMAPRO.2022.07.015.
53. A. S. Rajput, M. Das, and S. Kapil, "Characterization of wear resistance and corrosion during magnetorheological fluid assisted finishing (MFAF) of Ti-6Al-4V and duplex stainless steel for enhanced biocompatibility," *Proceedings of the Institution of Mechanical Engineers, Part E: Journal of Process Mechanical Engineering*, Jun. 2022, doi: 10.1177/09544089221107990.
54. A. K. Mondal, A. S. Rajput, D. Prasad, and D. Bose, *Magnetic Field Assisted Finishing Processes*, Springer, Cham, 2020, pp. 211–234.
55. K. Yung, S. Zhang, L. Duan, H. S. Choy, and Z. X. Cai, "Laser polishing of additive manufactured tool steel components using pulsed or continuous-wave lasers," *Int. J. Adv. Manuf. Technol.*, vol. 105, pp. 425–440, 2019.
56. E. Manco, E. Cozzolino, and A. Astarita, "Laser polishing of additively manufactured metal parts: A review," *Surf. Eng.*, vol. 38, no. 3, pp. 217–233, 2022, doi: 10.1080/02670844.2022.2072080.
57. K. C. Yung, T. Y. Xiao, H. S. Choy, W. J. Wang, and Z. X. Cai, "Laser polishing of additive manufactured CoCr alloy components with complex surface geometry," *J. Mater. Process. Technol.*, vol. 262, pp. 53–64, 2018.
58. A. Gisario, M. Barletta, and F. Veniali, "Laser polishing: A review of a constantly growing technology in the surface finishing of components made by additive manufacturing," *Int. J. Adv. Manuf. Technol.*, 120(3-4), 1433-1472, 2022.
59. S. Lee, Z. Ahmadi, J. W. Pegues, M. Mahjouri-Samani, and N. Shamsaei, "Laser polishing for improving fatigue performance of additive manufactured Ti-6Al-4V parts," *Opt. Laser Technol.*, vol. 134, p. 106639, Feb. 2021, doi: 10.1016/J.OPTLASTEC.2020.106639.
60. S. Park, P. Liu, K. Yi, G. Choi, K. Y. Jhang, and H. Sohn, "Mechanical properties estimation of additively manufactured metal components using femtosecond laser ultrasonics and laser polishing," *Int. J. Mach. Tools Manuf.*, 166, 103745, 2021.
61. T. Ermergen and F. Taylan, "Review on surface quality improvement of additively manufactured metals by laser polishing," *Arab. J. Sci. Eng.*, vol. 46, no. 8, pp. 7125–7141, Aug. 2021, doi: 10.1007/S13369-021-05658-9/FIGURES/20.
62. M. A. Obeidi, E. McCarthy, B. O'Connell, I. Ul Ahad, and D. Brabazon, "Laser polishing of additive manufactured 316L stainless steel synthesized by selective laser melting," *Materials*, vol. 12, no. 6, p. 991, 2019.
63. N. Li, P. Fan, Q. Zhu, B. Cui, J. F. Silvain, and Y. F. Lu, "Femtosecond laser polishing of additively manufactured parts at grazing incidence," *Appl. Surf. Sci.*, vol. 612, p. 155833, Mar. 2023, doi: 10.1016/J.APSUSC.2022.155833.
64. J. S. Chohan, R. Singh, and K. S. Boparai, "Vapor smoothing process for surface finishing of FDM replicas," *Mater. Today Proc.*, vol. 26, pp. 173–179, 2019, doi: 10.1016/j.matpr.2019.09.013.

65. X. Gong, T. Anderson, and K. Chou, "Review on powder-based electron beam additive manufacturing technology," In *International Symposium on Flexible Automation*, vol. 45110, pp. 507–515, 2012, June. American Society of Mechanical Engineers.
66. Y. Zhong, L. Rännar, S. Wikman, A. Koptyug, L. Liu, D. Cui, and Z. Shen, "Additive manufacturing of ITER first wall panel parts by two approaches: Selective laser melting and electron beam melting," *Fusion Eng. Des.*, vol. 116, pp. 24–33, 2017.
67. G. Vastola, G. Zhang, Q. Pei, and Y. W. Zhang, "Modeling and control of remelting in high-energy beam additive manufacturing," *Addit. Manuf.*, vol. 7, pp. 57–63, 2015.
68. N. Shen and K. Chou, "Thermal modeling of electron beam additive manufacturing process: powder sintering effects," In *International Manufacturing Science and Engineering Conference*, vol. 54990, pp. 287–295, 2012, American Society of Mechanical Engineers, doi: 10.1115/MSEC2012-7253.

9 Process automation for abrasive-based precision-finishing techniques

Saurabh Singh Rathore, Varun Sharma, Faiz Iqbal, Dilshad Ahmad Khan, Zafar Alam, and Navneet Arora

9.1 INTRODUCTION TO PROCESS AUTOMATION

The surface finishing of an object is essential for various reasons, such as for improving its appearance, increasing durability, enhancing functionality, and preventing corrosion. So, conventional finishing processes are developed for the finishing of objects, which play an essential role in achieving the desired level of finish on an object's surface. These processes typically involve using traditional techniques and equipment, such as sanding, polishing, buffing, grinding, honing, and lapping, to alter and improve the surface of an object. However, precision finishing is challenging with traditional finishing techniques because the cutting edge's relative motion to the workpiece results in unpredictable forces that damage the subsurface. So, advanced nano-finishing processes needs to be developed for precise finishing with dimensional accuracy.

The Fourth Industrial Revolution (Industry 4.0) emphasizes new technologies such as automation, robotics, and Artificial Intelligence (AI) to create more efficient and effective production processes. In precision finishing, Industry 4.0 has the potential to revolutionize the way objects are finished by enabling more precise, accurate, and consistent finishing techniques. This chapter explores several abrasive-based finishing techniques and describes the process parameters.

Process automation uses technology to automate repetitive or manual processes to improve efficiency, reduce costs, and increase the productivity. The benefits of process automation are shown in Figure 9.1. Software applications, robots, AI, and other advanced technologies can achieve this process automation in precision finishing to ensure consistent quality and accuracy. Another benefit of process automation in abrasive-based precision finishing is that this can reduce the amount of time and labour required to complete the task.

DOI: 10.1201/9781003496298-9

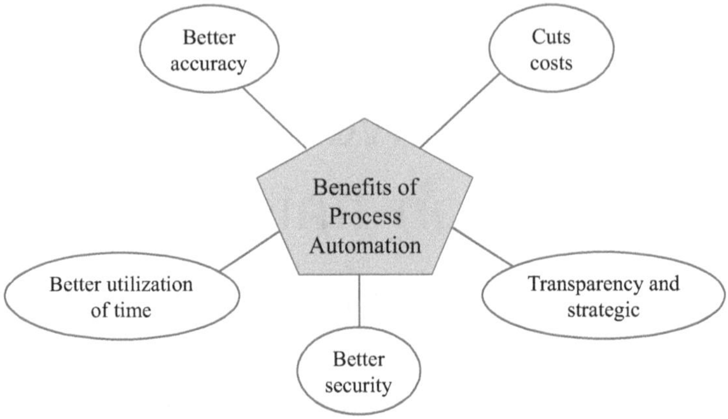

FIGURE 9.1 Benefits of process automation.

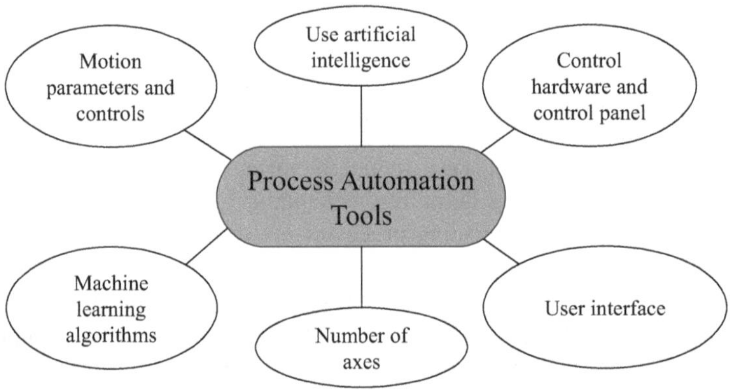

FIGURE 9.2 Process automation tools.

9.2 PROCESS AUTOMATION TOOLS

The process automation tools can help abrasive-based precision finishing to achieve greater efficiency, accuracy, and consistency in their finishing processes while reducing the risk of error and waste. The tools of process automation are shown in Figure 9.2.

9.2.1 Motion parameters and control

Motion parameters and control are essential for creating realistic and accurate movement in various applications and the movement of physical objects. Each abrasive-based finishing process requires the control of motion parameters to control the workpiece and the tool movement properly. There exist multiple ways of providing the power to the worktable and tool.

A motion control system controls a machine's or mechanical system's movement, such as a robot, CNC (Computer Numerical Control) machine, or production line

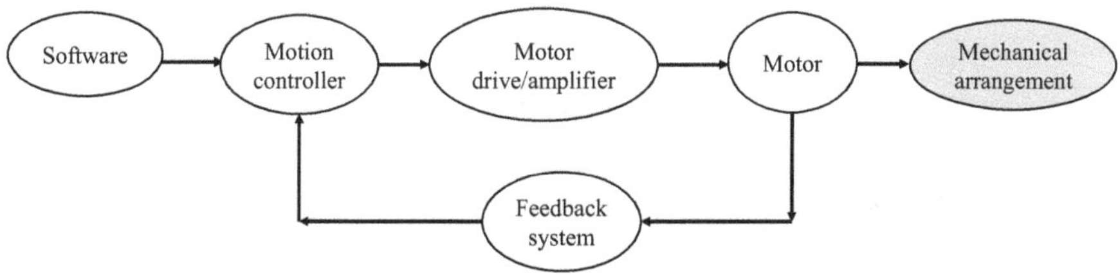

FIGURE 9.3 Motion control system.

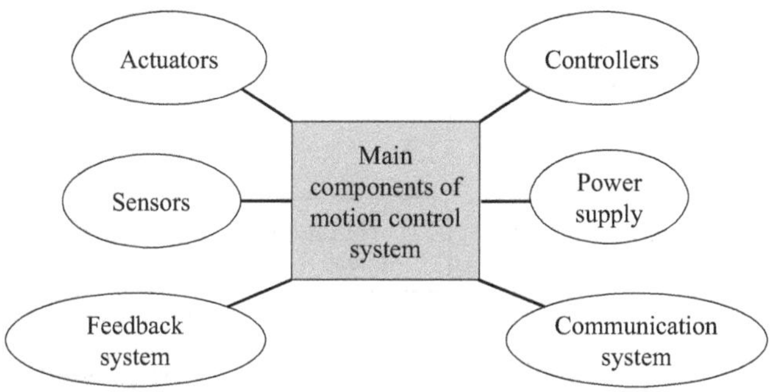

FIGURE 9.4 Main components of the motion control system.

(Figure 9.3). The primary function of a motion control system is to generate and control motion in an accurate and repeatable manner; for this, various components of the motion control system are required, which are shown in Figure 9.4.

9.2.2 Actuators

Actuators, typically motors, are used in abrasive-based finishing techniques to impart motion to the worktable or tool. The actuators generate activity in the system, such as linear or rotary motion. Actuators convert an input signal, typically electrical, hydraulic, or pneumatic, into mechanical movement or force. They are used in motion control systems to generate motion in a specific direction, speed, and power to accomplish a particular task. Actuators come in different varieties (Figure 9.5) and selecting the best actuator for a given application depends on the system's needs.

- **Electric actuators**

  Electric actuators are powered by electricity and convert electrical energy into mechanical motion. AC or DC power sources can power them and are often used in applications that require high precision and accuracy. The different types of electric actuators are shown in Figure 9.6.
- **Hydraulic actuators**

  Hydraulic actuators use oil or water to produce mechanical motion. As a result of their high force output, they are commonly used in heavy-duty applications.

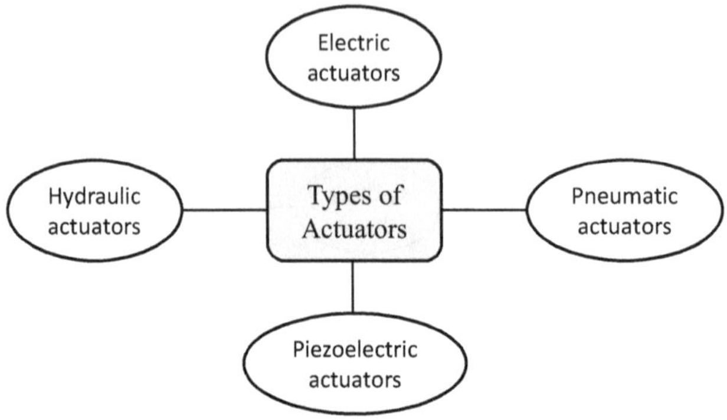

FIGURE 9.5 Types of actuators.

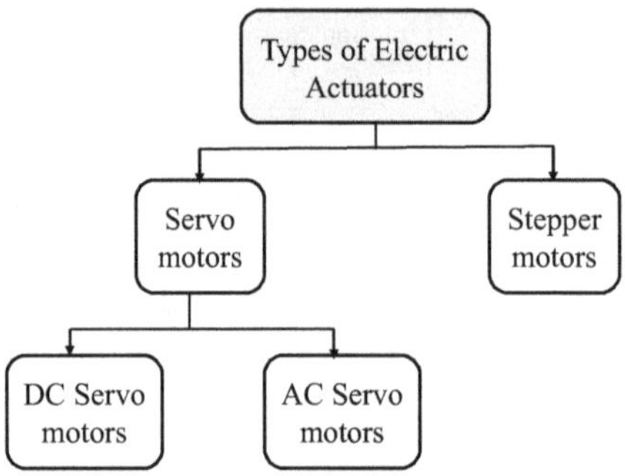

FIGURE 9.6 Types of electric actuators.

- **Pneumatic actuators**
  Pneumatic actuators produce mechanical motion by using compressed air. They are known for their fast response time and high-force output and are commonly used in industrial applications.
- **Piezoelectric actuators**
  A piezoelectric actuator converts electrical energy into mechanical motion using a piezoelectric material. They are known for their fast response time and high precision, and they are commonly used in applications that require fine-tuned control.

9.2.3 Controllers

The controllers are the central part of the motion control system. They receive inputs from sensors and other sources and use algorithms to calculate appropriate output to send to the actuators. The examples of controllers include PLCs (Programmable Logic Controllers), CNCs, and motion controllers.

9.2.4 Sensors

The sensors provide feedback to the controllers about the system's position, velocity, and other parameters. Sensors play a critical role in abrasive-based precision finishing processes by providing real-time data and feedback of various parameters such as force, pressure, temperature, and vibration. The examples of sensors include encoders, linear and rotary sensors, and accelerometers. Sensors play a crucial role in modern-day machines. They detect and measure various physical, chemical, or biological parameters to enable devices to operate more efficiently, accurately, and safely. Some of the most common sensors used in machines are shown in Figure 9.7.

9.2.5 Power supply

The power supply provides the necessary power to the actuators to generate motion in the system. Power supply in machines is critical for their operation, as it allows for the required electrical energy to power the machine's components and subsystems.

9.2.6 Feedback system

A feedback system can be helpful in abrasive-based precision finishing processes. The feedback system allows the process to be continuously monitored and adjusted to achieve desired results. The feedback system provides position, velocity, and other aspects of the system to the controller. This information is used to adjust the system's

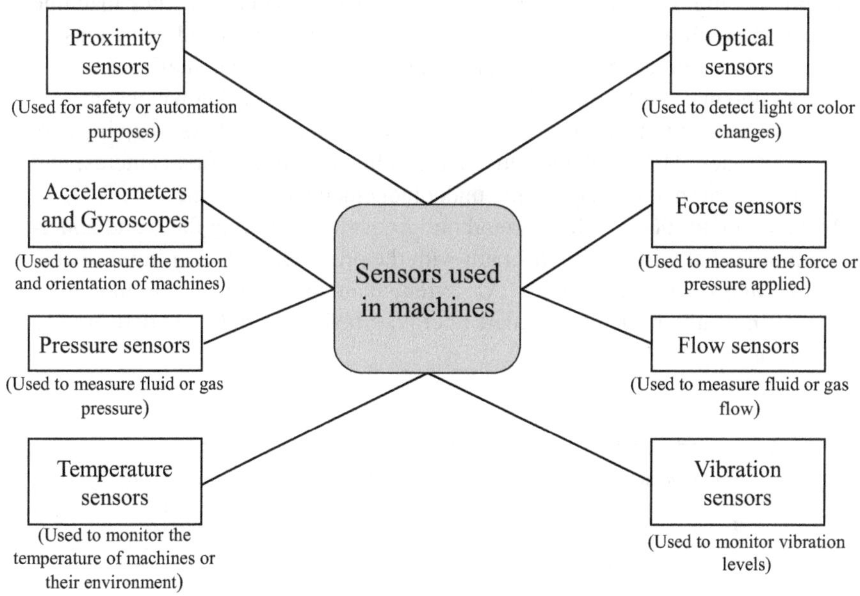

FIGURE 9.7 Sensors used in machines.

motion. In a feedback system, the production of a procedure or method is measured and compared to a desired or reference value. The resulting error is used to adjust the inputs to the operation or system to reduce the error. This is done to bring the output closer to the desired value.

9.2.7 Communication system

A communication system enables the exchange of information between two or more entities, such as computers or devices. The communication system is responsible for transferring the data between different motion control system components, including controllers, sensors, and actuators. The communication systems can be analog or digital and can operate over various media types, including wire, radio waves, optical fibres, and satellite links. The flow diagram of the communication system is shown in Figure 9.8.

9.2.8 Artificial intelligence

AI creates computer systems capable of doing tasks that usually require human intelligence. These abilities include spoken language processing, speech recognition, visual perception, and decision-making [1]. AI systems use a combination of algorithms, data, and Machine Learning (ML) techniques to learn from experience and improve their performance over time. AI has many machine applications, from simple devices to complex industrial systems. A few examples of integration of AI in machines are listed below:

To prevent breakdowns and optimize performance, AI can monitor machine conditions in real time. By examining information gathered from sensors and other sources, AI systems can identify patterns and anomalies that indicate potential problems and alert operators before a breakdown occurs.

AI is widely used in developing autonomous vehicles, such as self-driving cars and drones. These vehicles utilize a combination of sensors, cameras, and ML algorithms to navigate and make judgements in real time.

AI can track product quality throughout production, spotting flaws and abnormalities that might point to a fault with the procedure. By examining sensor data and other data sources, AI systems can spot patterns and anomalies that indicate problems and alert operators before a product is delivered to customers.

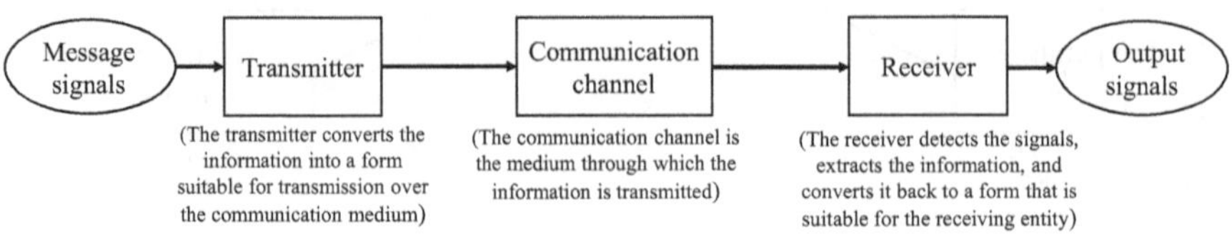

FIGURE 9.8 Communication system.

AI is utilized extensively in robotics, enabling robots to execute complicated tasks in diverse environments, such as manufacturing, logistics, and healthcare. Using ML algorithms, robots can adapt changing situations and learn from experience to improve their performance.

9.2.9 Control hardware and control panel

Utilizing control hardware and the control panel is a crucial aspect of process automation that should be considered when automating magnetic field-assisted finishing processes. Control hardware and control panels are both components of a control system that enable the monitoring and controlling industrial processes.

Control hardware: Control hardware refers to the physical components of a control system that enable communication and control between the controller and the process. Control hardware typically includes sensors, actuators, PLCs, and other parts, as shown in Figure 9.9. Sensors detect changes in the process, such as temperature, pressure, or flow rate, and send that information to the controller. The controller uses that information to make decisions and send commands to the actuators, which control the process by adjusting valves, pumps, motors, or other equipment. PLCs coordinate the communication between the sensors, controller, and actuators.

Control panel: A control panel is a user interface that allows operators to monitor and control the industrial process. Control panels typically include displays, switches, buttons, and other controls that enable operators to view process variables and adjust setpoints or control parameters.

9.2.10 ML algorithms

ML algorithms are progressively employed in machines to facilitate automation and improve performance. Here are a few examples of how ML algorithms are being used in devices:

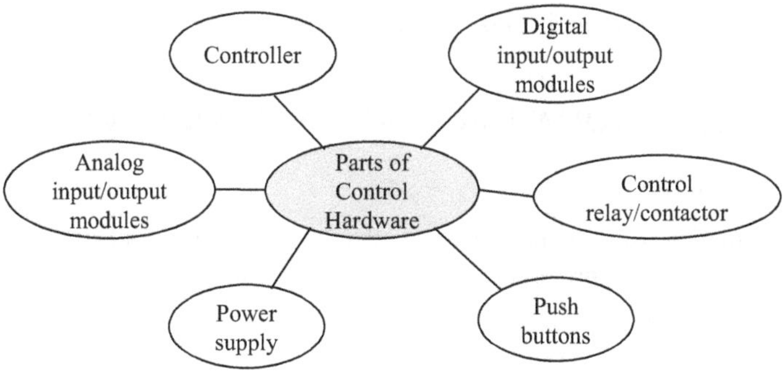

FIGURE 9.9 Parts of control hardware.

ML algorithms are used in computer vision applications to recognize and classify images. For example, ML algorithms can detect product defects on a production line by analysing camera images.

ML algorithms can optimize the performance of machines by adjusting parameters and control settings in real time. For example, ML algorithms can be employed to adjust the speed and torque of motors in response to changing load conditions.

9.2.11 Number of Axes

The number of axes a machine can control depends on the specific device and its intended application. However, many modern machines can control multiple axes simultaneously, enabling complex movements and operations to be performed with high precision.

For example, a CNC machine typically has three or more axes, which control the movement of a cutting tool or other workpiece. As the workpiece's horizontal, vertical, and depth dimensions are indicated by the three axes, the three axes are *X*, *Y*, and *Z*. Some CNC machines also have additional axes, such as a fourth axis that enables rotational movement or a fifth axis that allows tilting the workpiece. In addition to CNC machines, other machines may have multiple axes for specific applications. For example, a robotic arm may have multiple axes of movement to enable it to perform tasks such as picking up and moving objects in a factory or laboratory setting.

9.2.12 User Interface

When automating a mechanical system to a significant level, it is essential to include a user interface. User interfaces are created to see how an automated system communicates with the individual employing the system. A typical user interface operates in two modes:

Automatic mode: The user interface's automated mode sets the system's parameters before a process starts. After pressing the start button, the procedure then begins automatically.

Manual mode: A machine is operated directly by the operator without automation or computer control. The operator operates the equipment manually and often uses physical controls like levers, buttons, or pedals.

9.3 PROCESS AUTOMATION ADVANTAGES AND LIMITATIONS

9.3.1 Advantages

Process automation uses technology and software to automate and streamline manufacturing or production processes. Some of the advantages of process automation include the following:

Automated systems perform tasks faster and more consistently than human workers, reducing cycle times and increasing throughput.

Automation helps to improve production process accuracy and consistency.
Automation helps to improve safety in manufacturing and production environments.
Manufacturers can reduce their dependency on human labour by automating tasks, which could result in lower labour costs and increased profitability.

9.3.2 Limitations

While process automation offers many advantages, there are also limitations to its use. Some of the limitations of process automation include the following:

Implementing process automation systems can be expensive, requiring significant hardware, software, and training investment.
Automation systems often require technical design, implementation, and maintenance skills.
Process automation systems rely on technology and software, which can be vulnerable to failures, bugs, or cyber-attacks. This can lead to downtime and potential disruptions in production.
Automation can reduce human interaction in production processes, which can have social and psychological impacts on workers. In addition, automated systems may lack human workers' intuition and problem-solving skills, making it difficult to address specific issues or unexpected events.

9.4 ABRASIVE-BASED PRECISION FINISHING TECHNIQUES

Abrasive-based precision finishing techniques are surface-finishing methods that involve using abrasive particles to remove small amounts of material from a workpiece, to improve its surface quality and dimensional accuracy. Abrasive-based super finishing processes used for flat and freeform surfaces are:

- Magnetorheological Finishing (MRF) process
- Magnetic Abrasive Finishing (MAF) process
- Abrasive Flow Machining (AFM) process

9.4.1 MRF process

The MRF method uses a magnetically controlled fluid known as Magnetorheological Polishing (MRP) fluid to finish complex geometries and freeform surfaces precisely. Under magnetic field influence, the rheological characteristics of MRP fluids change and show a solid viscoelastic texture. The basic abrasive-based MRF precision finishing process is related to finishing flat surfaces. For automation consideration, we will focus on MRF variants and their parameters. Some variants of MRF are:

- Magnetorheological Abrasive Flow Finishing (MRAFF)
- Magnetorheological Abrasive Honing (MRAH)

- Ball End Magnetorheological Finishing (BEMRF)
- Chemo-Mechanical Magnetorheological Finishing (CM-MRF)

9.4.1.1 Process parameters for MRF

Process parameters refer to the factors in a particular manufacturing or production process. The fishbone diagram of the MRF process parameters is shown in Figure 9.10.

9.4.1.2 Process automation for MRF

Process automation can play a crucial role in MRF by improving the consistency and accuracy of the finishing process. Now we will study process automation in MRF variants to control their process parameters.

- MRAFF
  Two hydraulic cylinders move in and out to push and pull the MRP fluid over the workpiece during the MRAFF procedure (Figure 9.11). At the same time, the MRP fluid is subject to the influence of an energized electromagnet outside the polishing chamber. Automation can enable real-time monitoring of the MRAFF process to provide immediate feedback on the quality of the finished components, allowing for adjustments and corrective actions if required.
- MRAH
  Figure 9.12 shows the schematic of the MRAH process. In this, the workpiece is attached to the spindle and is given rotation on its axis; the only other motion in MRAH is the movement of a plunger that pushes and pulls the MR fluid to and fro over the workpiece. This magnetic field-assisted finishing process uses servo or stepper motors to move the polishing fluid plunger.
- BEMRF
  Figure 9.13 shows a schematic diagram of the BEMRF process. Pressurized MRP fluid is pumped into the hollow spindle during the BEMRF process.

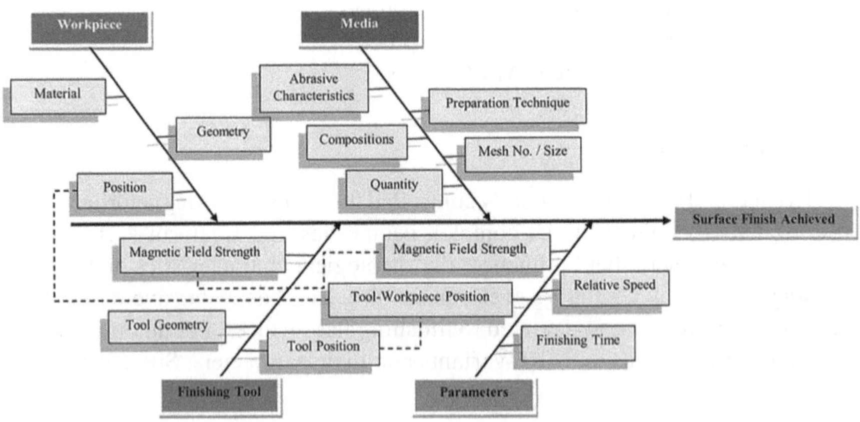

FIGURE 9.10 Fishbone diagram of the MRF process parameters (from Ahmad, S., Singari, R. M., & Mishra, R. S., Taylor & Francis, 98(4), 186–198, 2020) [2].

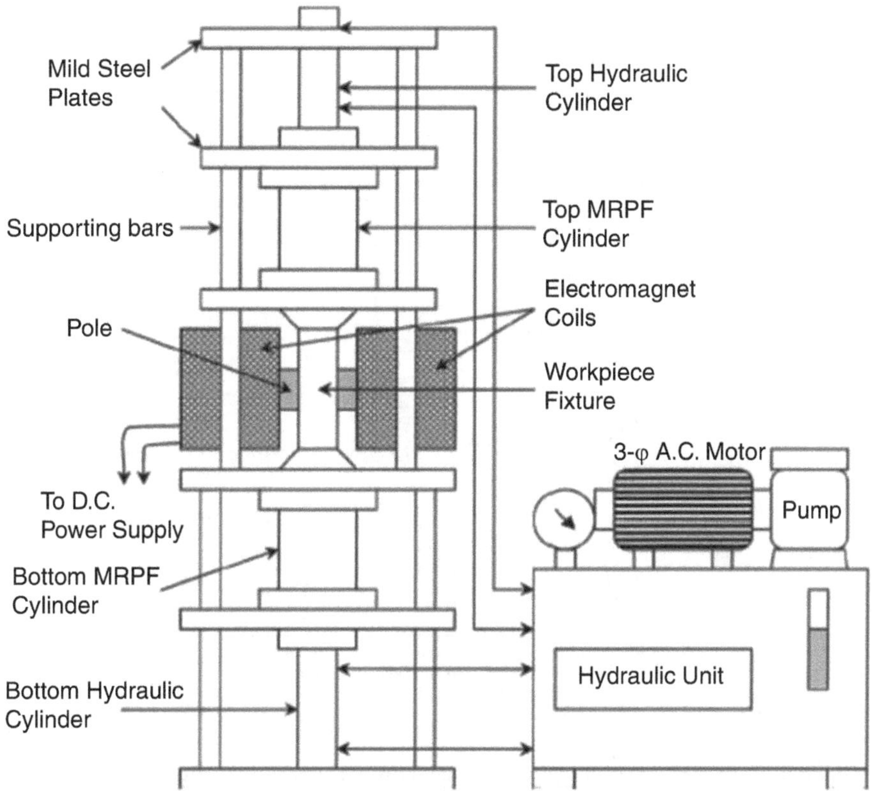

FIGURE 9.11 Schematic of the MRAFF process (from Jha, S., & Jain, V. K., *International Journal of Machine Tools and Manufacture*, Elsevier, 44(10), 1019–1029, 2004) [3].

An electric motor turns the entire assembly while the spindle is inside an electromagnet. This arrangement is used for flat workpiece surfaces and replica profiles of the tooltip. Still, 3D freeform surfaces are finished using process atomization and named a five-axis CNC BEMRF process. There have been various advancements in the development of BEMRF process over the last decade [5–10]

The five-axis CNC BEMRF machine can move in three linear directions along the *X*, *Y*, and *Z* axes and two rotating directions along the *B* and *C* axes, respectively. The *B* and *C* rotation axes are orthogonal. *B* and *C* are parallel to *Y* and *Z*, respectively. The *X*, *Y*, and *Z* linear axes are orthogonal. Figure 9.14 shows the BEMRF setup arrangement of five motion axes and the BEMRF tool head.

- CM-MRF process

Figure 9.15 shows the schematic diagram of the CM-MRF setup. CM-MRF [13] is an advanced polishing process to finish precision optical surfaces such as lenses, mirrors, and prisms. This process combines chemical reactions, mechanical abrasion, and magnetorheological (MR) fluid technology to

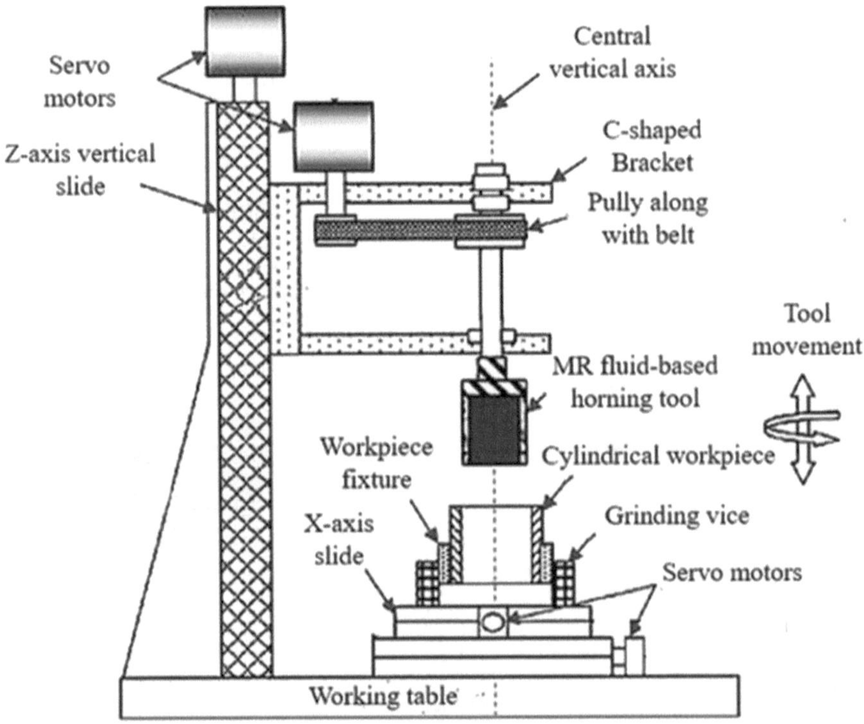

FIGURE 9.12 Schematic of the MRAH process (from Paswan, S. K., Bedi, T. S., & Singh, A. K., Elsevier, 376, 1207–1221, 2017) [4].

produce high-quality surface finishes. Process automation in the CM-MRF process involves a controller, three-axis CNC-based machine, a slurry system, a finishing head, and a working table adjustment by using various computer software and hardware systems to automate and control the different stages of the process.

9.4.2 MAF Process

MAF is a surface-finishing process that uses a magnetic field to control the movement of abrasive particles to remove the material from a workpiece. This process achieves a high-quality surface finish on various materials, including metals, ceramics, and composites. Process automation in the MAF process involves using multiple computer software and hardware systems to automate and control the different stages of the process. Some variants of MAF are as follows:

- Internal finishing by magnetic field-assisted finishing
- Electrolytic Magnetic Abrasive Finishing (EMAF)
- Cylindrical Magnetic Abrasive Finishing (CMAF)
- Ultrasonic-Assisted Magnetic Abrasive Finishing (UAMAF)

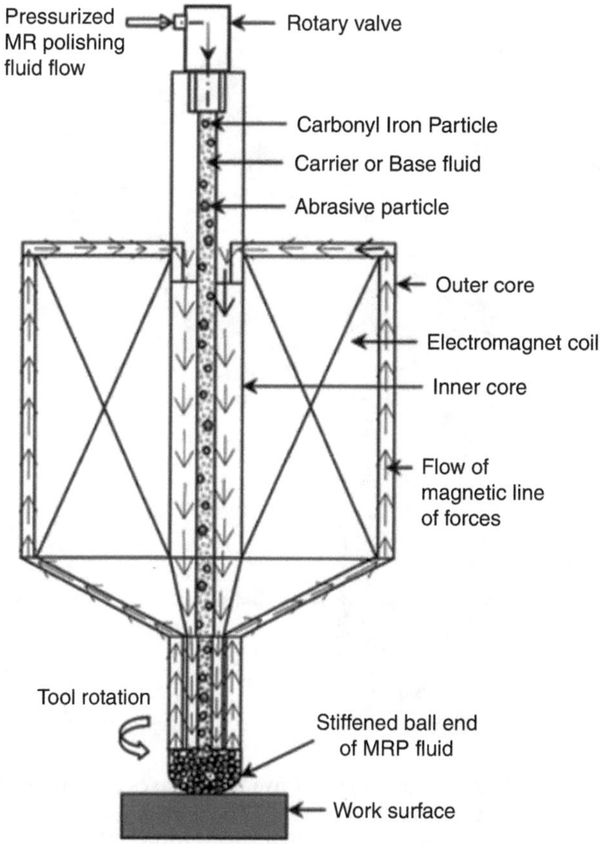

FIGURE 9.13 Schematic of the BEMRF process (from Singh, A. K., Jha, S., & Pandey, P. M., *Materials and Manufacturing Processes*, Taylor & Francis, 27(4), 389–394, 2012) [11].

9.4.2.1 Process parameters for MAF

Process parameters are essential to achieve the appropriate surface smoothness and material removal rate during MAF. MAF process parameters are like those of MRF. The main difference between MRF and MAF is the abrasive particles used. MRF is a process that uses a slurry containing magnetorheological fluid, which is a type of intelligent fluid that changes its viscosity in response to a magnetic field. MAF is a process that uses abrasive particles that are magnetically susceptible, such as iron oxide particles, which are mixed with a non-magnetic carrier fluid to form a slurry. The fishbone diagram of the MAF process parameters is shown in Figure 9.16.

9.4.2.2 Process automation for MAF

Process automation for MAF involves using advanced control systems and robotics to automate various aspects of the finishing process. This can improve the finishing process's consistency, accuracy, and efficiency, reducing the need for manual intervention and operator skills. Now we study process automation in MAF variants to control their process parameters.

FIGURE 9.14 BEMRF setup arrangement of five motion axes and the BEMRF tool head (from Alam, Z., Iqbal, F., Ganesan, S., & Jha, S. *The International Journal of Advanced Manufacturing Technology*, Springer, 100(5), 1031–1042, 2019) [12].

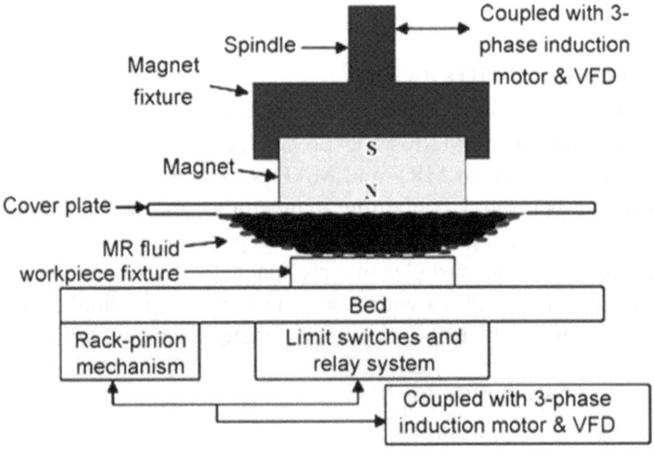

FIGURE 9.15 Schematic diagram of the CM-MRF setup (from Jain, V. K., Ranjan, P., Suri, V. K., & Komanduri, R., *CIRP Annals*, Elsevier, 59(1), 323–328, 2010) [13].

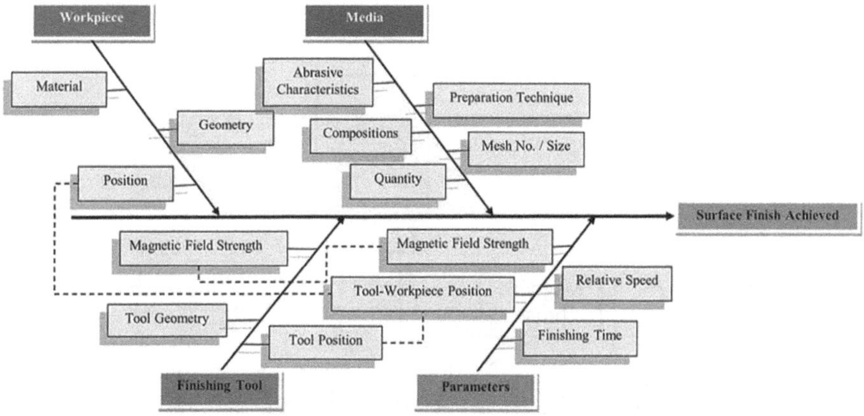

FIGURE 9.16 Fishbone diagram of the MAF process parameters (from Ahmad, S., Singari, R. M., & Mishra, R. S., Taylor & Francis, 98(4), 186–198, 2020) [2].

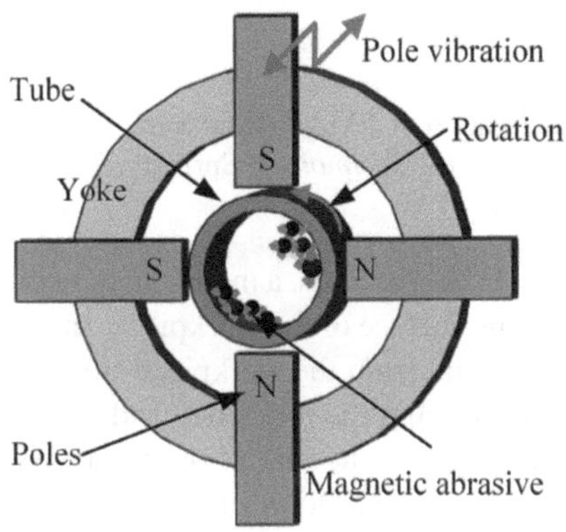

FIGURE 9.17 Schematic diagram of internal finishing by MAF (from Yamaguchi, H., & Shinmura, T., *Precision Engineering*, Elsevier, 28(2), 135–142, 2004) [14].

- Internal finishing by magnetic field-assisted finishing
  Internal finishing by MAF is used to finish interior surfaces of complex, difficult-to-access parts such as bores, cylinders, and pipes. Process automation can enhance the effectiveness and efficiency of internal finishing through MAF. Process automation in MAF uses robotic systems to control the positioning of the workpiece and tool. The robot can be programmed to move the workpiece precisely and accurately along the magnetic tool, ensuring uniform and consistent material removal and polishing using advanced control systems to regulate the slurry's flow rate and the workpiece's movement (Figure 9.17).
- EMAF
  The schematic diagram of the EMAF process is shown in Figure 9.18. EMAF is a hybrid finishing process that combines electrolytic machining

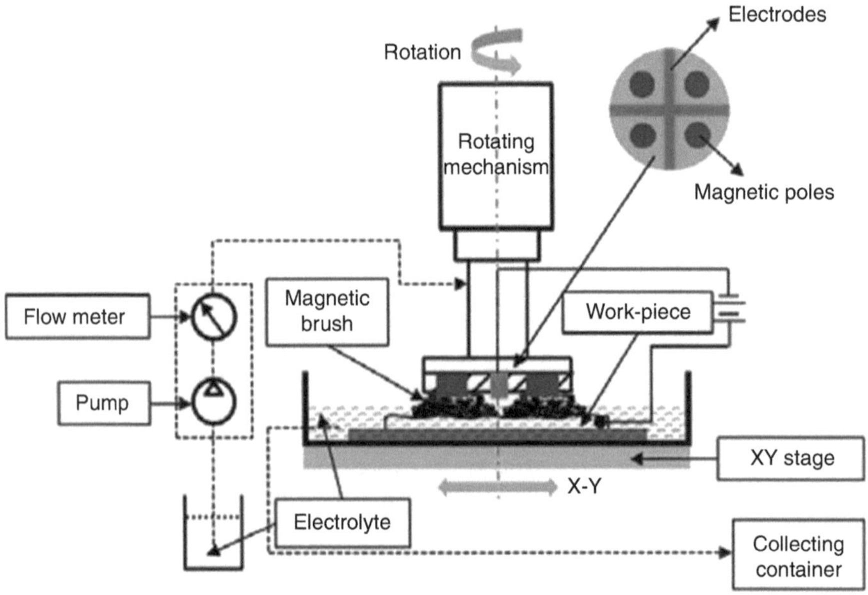

FIGURE 9.18 Schematic diagram of EMAF (from Sun, X., & Zou, Y., *The International Journal of Advanced Manufacturing Technology*, Springer, 92, 3373–3384, 2017) [15].

and MAF principles. In this process, a magnetic field is used to hold abrasive particles in place on the surface of the workpiece. In contrast, an electrolytic solution dissolves material from the workpiece surface. Process automation can be essential in improving the efficiency and consistency of EMAF. An automated system can be used to control the electrolyte solution's flow and monitor and adjust the concentration of the solution as needed. This can help to maintain consistent finishing conditions and improve the quality of the finished product. Automated sensors and control systems can monitor and control various process parameters.

- CMAF

  Figure 9.19 shows the schematic diagram of a CMAF. CMAF is a surface-finishing process used to improve the surface quality of cylindrical components. Process automation of CMAF can significantly enhance the efficiency and consistency of the process. Automation can help to reduce operator error, increase productivity, and improve the quality of finished parts. The workpiece can be placed on a rotating fixture or conveyor belt to move it through the process. Automated systems can monitor the process parameters, such as rotational speed, magnetic field strength, and abrasive flow rate, to ensure consistent finishing results.
- UAMAF

  UAMAF is an advanced surface-finishing process that combines the benefits of ultrasonic vibrations with MAF. This process is particularly effective for finishing hard and brittle materials such as ceramics, glass, and hardened steel. Process automation of UAMAF can further improve the

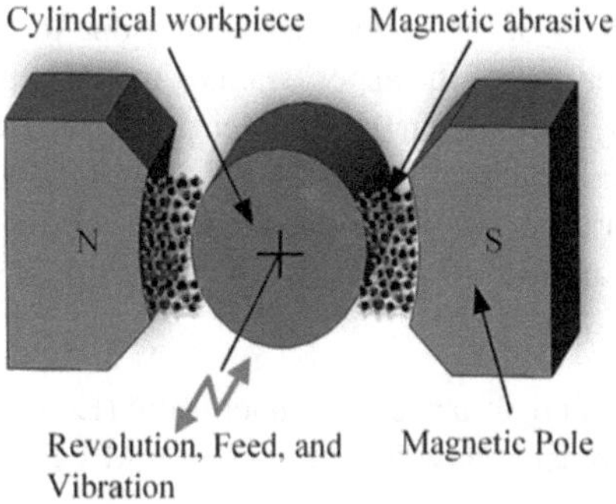

FIGURE 9.19 Schematic diagram of cylindrical MAF (from Qian, C., Fan, Z., Tian, Y., Liu, Y., Han, J., & Wang, J., *The International Journal of Advanced Manufacturing Technology*, Springer, 112, 619–634, 2020) [16].

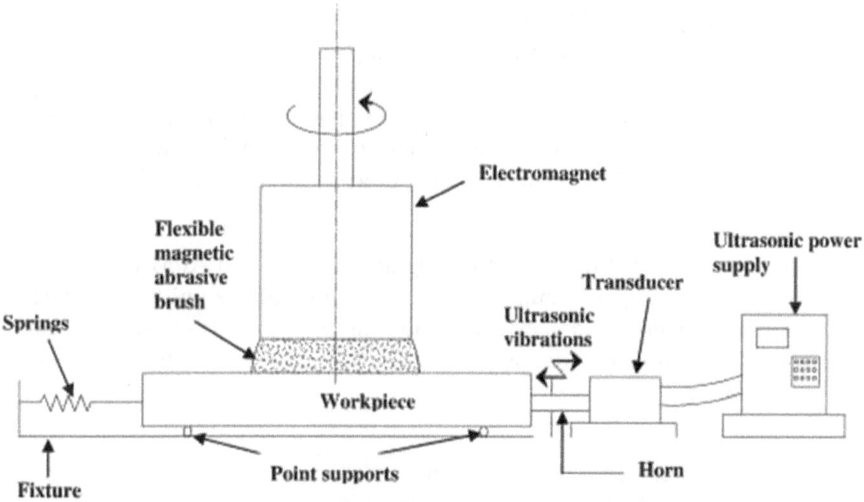

FIGURE 9.20 Schematic diagram of UAMAF (from Mulik, R. S., & Pandey, P. M., *International Journal of Refractory Metals and Hard Materials*, Elsevier, 29(1), 68–77, 2011) [17].

efficiency and consistency of the process. The ultrasonic transducer can be controlled using an automated system to ensure consistent and precise vibrations. This can be achieved using feedback control systems. Computerized systems can monitor the process parameters such as rotational speed, magnetic field strength, abrasive flow rate, and ultrasonic vibration frequency and amplitude to ensure consistent finishing results (Figure 9.20).

9.4.3 AFM PROCESS

Abrasive-laden media are extruded across or through the restricted passage between the workpiece and tooling. Abrasion of media occurs due to extrusion, resulting in a finishing surface. Finishing of the surface will occur due to elastic deformation correlated with rubbing, shearing of the asperities, i.e., micro cutting, and flattening due to plastic deformation by moving particles under pressure. The basic abrasive-based AFM precision finishing process is related to finishing internal cylindrical surfaces. For automation consideration, we will focus on AFM variants and their parameters. Some variants of AFM are as follows:

- MRAFF
- Electrochemical-aided abrasive flow machining (ECAAFM)
- Drill-bit-guided abrasive flow finishing (DBG-AFF)

9.4.3.1 Process parameters for AFM

The process parameters are crucial to achieving the appropriate surface finish and material removal rate when using AFM. The fishbone diagram of the AFM process parameters is shown in Figure 9.21.

9.4.3.2 Process automation for AFM

Process automation in AFM involves using computer-controlled systems to control the process parameters and ensure consistency and accuracy in the machining process. The automation system controls the process parameters, monitors the process in real time, and adjusts to achieve the desired surface finish. Now we study process automation in AFM variants to manage their process parameters.

- AFM

  Figure 9.22 shows the schematic diagram of AFM. An AFM machine is used for deburring, polishing, and improving the surface finish of complex internal geometries. Process automation can use control software and sensors that control the process parameters such as pressure, flow rate, and temperature.

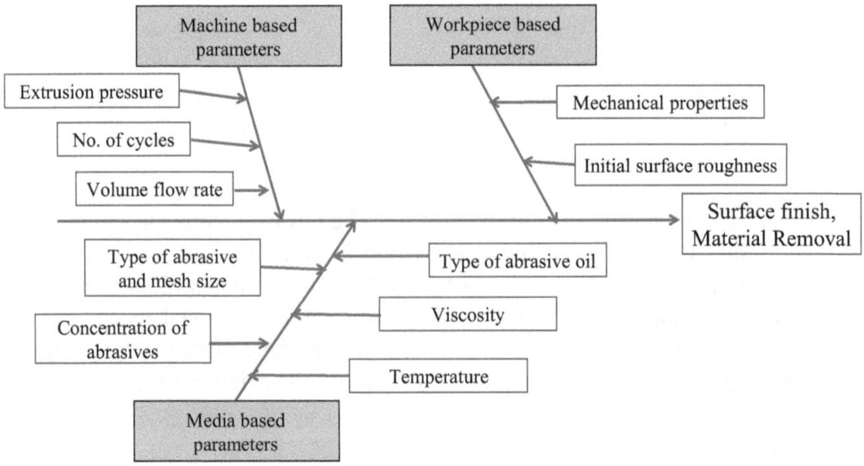

FIGURE 9.21 Fishbone diagram of the AFM process parameters.

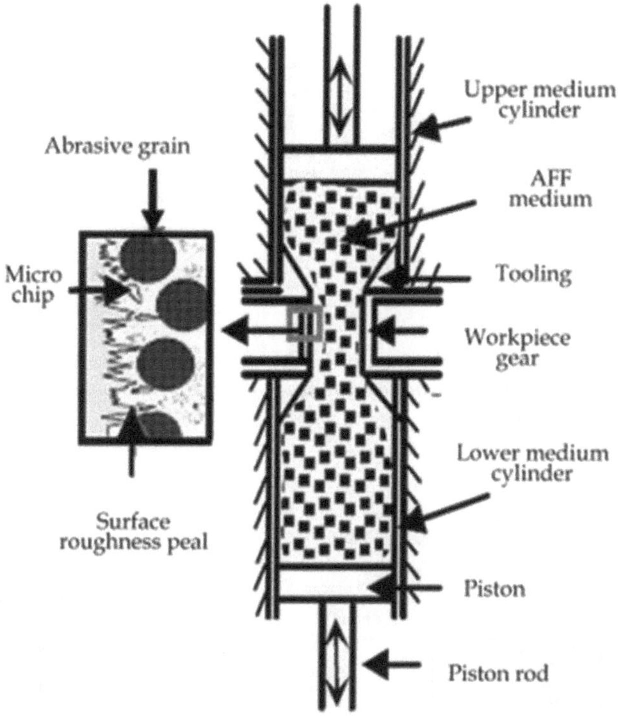

FIGURE 9.22 The schematic diagram of AFM (from Singh, S., Kumar, H., Kumar, S., & Chaitanya, S., *Materials Today Proceedings*, Elsevier, 56, 3108–3116,2022) [18].

In Abrasive Flow Finishing (AFF), when providing ultrasonic motion, it is called Ultrasonic-Assisted Abrasive Flow Machining (UAAFM) (Figure 9.23). Process automation in UAAFM can help to improve process repeatability. UAAFM [19] can be automated using in situ monitoring and AI. AI algorithms can be trained using data from previous machining operations to identify patterns and optimize the machining process.

- MRAFF

MRAFF is a hybrid machining process that combines the principles of AFM and MRF. In MRAFF, hydraulic pressure forces an MR fluid containing abrasive particles through the workpiece. At the same time, to control MR fluid flow, a magnetic field is applied. Process automation becomes increasingly important because of the characteristics necessary for adequate control. These characteristics include the management of hydraulics and pump and electromagnet functions (Figure 9.24).

The MRAFF process uses a motor to impart rotary motion to the magnet arrangement. This process is called Rotational Magnetorheological Abrasive Flow Finishing (R-MRAFF) (Figure 9.25). In R-MRAFF, a rotating magnetic field provides rotation to the MR polishing fluid medium. A hydraulic unit also offers reciprocating motion to the polishing medium.

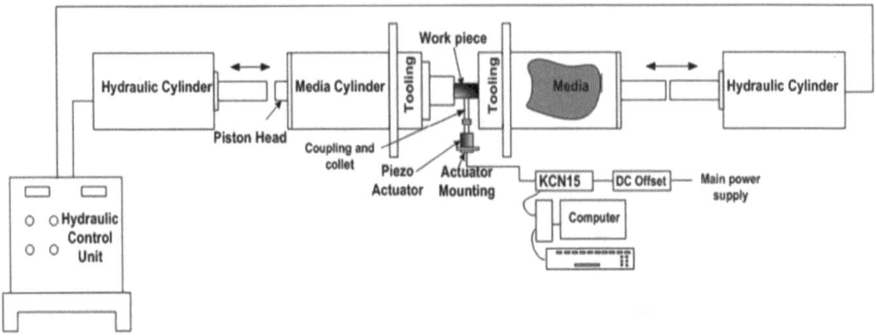

FIGURE 9.23 The schematic diagram of UAAFM (from Sharma, A. K., Venkatesh, G., Rajesha, S., & Kumar, P., *The International Journal of Advanced Manufacturing Technology*, Springer, 80, 477–493, 2015) [18].

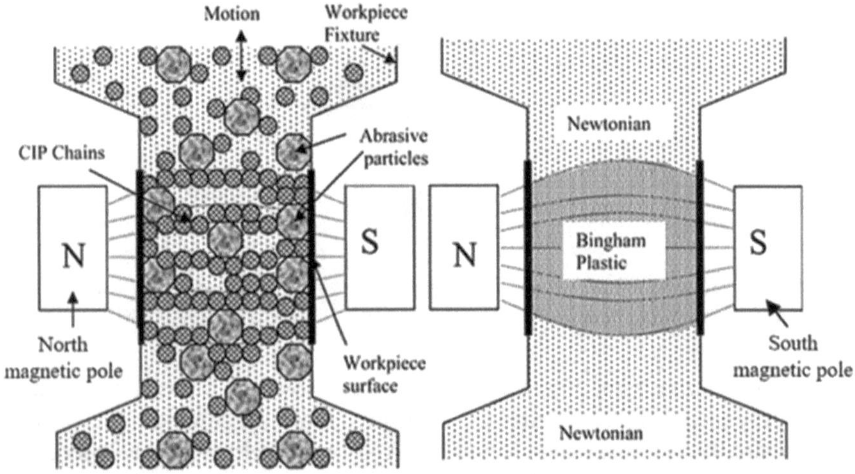

FIGURE 9.24 Magnetorheological abrasive flow finishing process (from Jha, S., & Jain, V. K., *International Journal of Machine Tools and Manufacture*, Elsevier, 44(10), 1019–1029, 2004) [3].

- ECAAFM

  ECAAFM (Figure 9.26) is a hybrid machining process that combines the principles of AFM and ECM. Process automation in ECAAFM controls the electric current and the electrolyte flow rate. Automated systems can use sensors and feedback control mechanisms to monitor and adjust real-time current and flow rates—process automation in ECAAFM to control the workpiece position and orientation. Automated systems can use robotics or computer-controlled mechanisms to position and orient the workpiece during machining precisely. This can help ensure that the machining process is performed accurately and consistently, even for complex geometries.

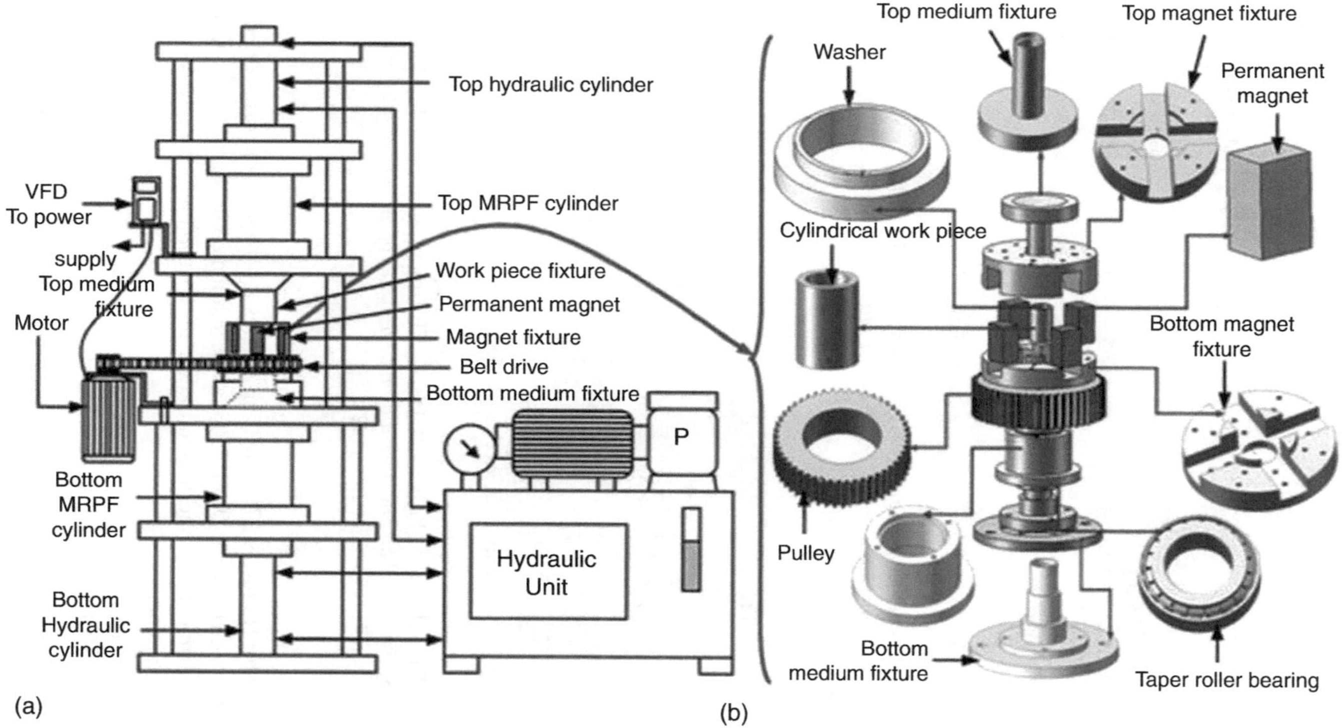

FIGURE 9.25 (a) Schematic of R-MRAFF process and (b) magnified view of the finishing zone consisting of the workpiece, workpiece fixture, and magnet fixture (from Das, M., Jain, V. K., & Ghoshdastidar, P. S., *Materials and Manufacturing Processes*Taylor & Francis, 26–8, 1073–1084, 2011) [20].

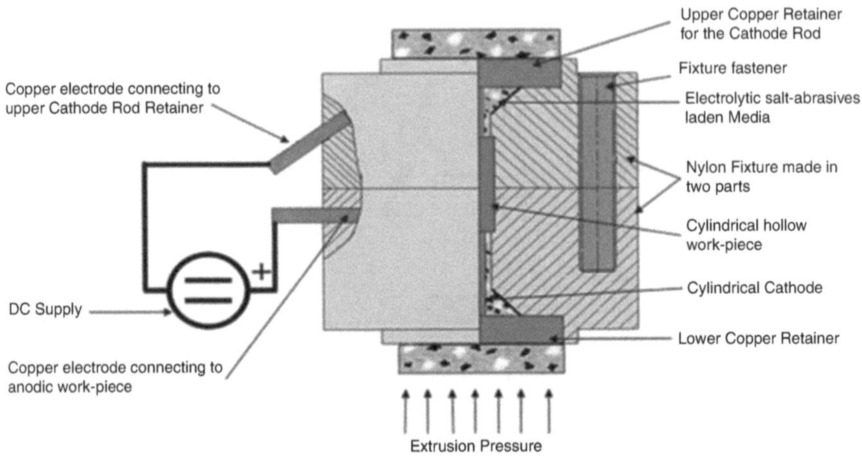

FIGURE 9.26 The schematic diagram of ECAAFM (from Brar, B. S., Walia, R. S., & Singh, V. P., *The International Journal of Advanced Manufacturing Technology*, Springer, 79, 329–342, 2015) [21].

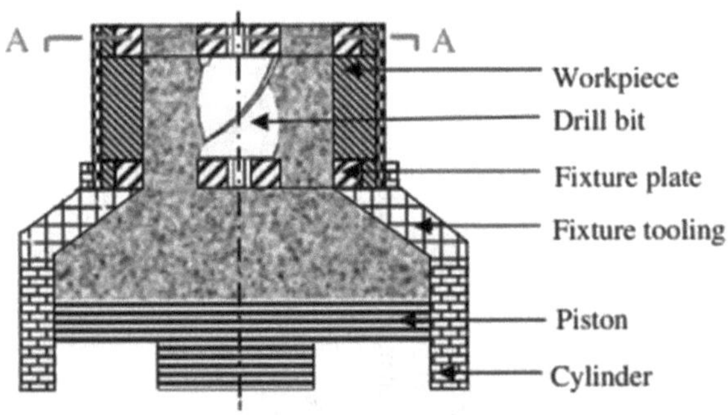

FIGURE 9.27 The schematic diagram of DBG-AFF (from Sankar, M. R., Mondal, S., Ramkumar, J., & Jain, V. K., *The International Journal of Advanced Manufacturing Technology*, Springer, 42, 678–688, 2009) [22].

- DBG-AFF

  In DBG-AFF (Figure 9.27), a drill bit creates a path for an abrasive media to flow through the part. DBG-AFF can be automated using robotic control. A robotic arm can be used to manipulate the drill bit and the interest in a controlled manner.

9.5 CONCLUSION

Many researchers have worked on abrasive-based precision finishing processes related to flat surfaces or specific techniques developed for specific surface profile finishing. Unique fixtures and process automation are required to finish surfaces other than flat surfaces; e.g., 3D profiles.

Process automation can bring significant benefits to abrasive-based superfinishing processes. By automating the process, manufacturers can achieve consistent and repeatable results, improve quality control, reduce variability and human error, and increase productivity and efficiency. Advanced technologies such as sensors, control systems, and robotics can enable real-time monitoring and adjustments to the process, leading to higher accuracy and precision.

REFERENCES

1. Joiner, I. A. (2018). *Emerging library technologies: it's not just for geeks*. Chandos Publishing. 9 August 2018.
2. Ahmad, S., Singari, R. M., & Mishra, R. S. (2020). Modelling and optimisation of magnetic abrasive finishing process based on a non-orthogonal array with ANN-GA approach. *Transactions of the IMF*, 98(4), 186–198.
3. Jha, S., & Jain, V. K. (2004). Design and development of the magnetorheological abrasive flow finishing (MRAFF) process. *International Journal of Machine Tools and Manufacture*, 44(10), 1019–1029.
4. Paswan, S. K., Bedi, T. S., & Singh, A. K. (2017). Modelling and simulation of surface roughness in magnetorheological fluid based honing process. *Wear*, 376, 1207–1221.
5. Khan, D. A., Alam, Z., Iqbal, F., & Jha, S. (2017, July). A study on the effect of polishing fluid composition in ball end magnetorheological finishing of aluminium. In *39th International MATADOR Conference on Advanced Manufacturing*, Manchester.
6. Iqbal, F., Alam, Z., & Jha, S. (2020). Modelling of transient behaviour of roughness reduction in ball end magnetorheological finishing process. *International Journal of Abrasive Technology*, 10(3), 170–192.
7. Iqbal, F., Alam, Z., Khan, D. A., & Jha, S. (2022). Automated insular surface finishing by ball end magnetorheological finishing process. *Materials and Manufacturing Processes*, 37(4), 437–447.
8. Iqbal, F., & Jha, S. (2016). Nanofinishing of free form surfaces using BEMRF. In V. K. Jain, *Nanofinishing Science and Technology* (pp. 255–284). CRC Press.
9. Alam, Z., Khan, D. A., Iqbal, F., & Jha, S. (2019). Effect of polishing fluid composition on forces in ball end magnetorheological finishing process. *International Journal of Precision Technology*, 8(2–4), 365–378.
10. Iqbal, F., & Jha, S. (2019). Experimental investigations into transient roughness reduction in ball-end magneto-rheological finishing process. *Materials and Manufacturing Processes*, 34(2), 224–231.
11. Singh, A. K., Jha, S., & Pandey, P. M. (2012). Magnetorheological ball end finishing process. *Materials and Manufacturing Processes*, 27(4), 389–394.
12. Alam, Z., Iqbal, F., Ganesan, S., & Jha, S. (2019). Nanofinishing of 3D surfaces by automated five-axis CNC ball end magnetorheological finishing machine using customized controller. *The International Journal of Advanced Manufacturing Technology*, 100, 1031–1042.
13. Jain, V. K., Ranjan, P., Suri, V. K., & Komanduri, R. (2010). Chemo-mechanical magneto-rheological finishing (CMMRF) of silicon for microelectronics applications. *CIRP Annals*, 59(1), 323–328.
14. Yamaguchi, H., & Shinmura, T. (2004). Internal finishing process for alumina ceramic components by a magnetic field assisted finishing process. *Precision Engineering*, 28(2), 135–142.
15. Sun, X., & Zou, Y. (2017). Development of magnetic abrasive finishing combined with electrolytic process for finishing SUS304 stainless steel plane. *The International Journal of Advanced Manufacturing Technology*, 92, 3373–3384.

16. Qian, C., Fan, Z., Tian, Y., Liu, Y., Han, J., & Wang, J. (2021). A review on magnetic abrasive finishing. *The International Journal of Advanced Manufacturing Technology*, 112, 619–634.
17. Mulik, R. S., & Pandey, P. M. (2011). Ultrasonic assisted magnetic abrasive finishing of hardened AISI 52100 steels using unbonded SiC abrasives. *International Journal of Refractory Metals and Hard Materials*, 29(1), 68–77.
18. Singh, S., Kumar, H., Kumar, S., & Chaitanya, S. (2022). A systematic review on recent advancements in Abrasive Flow Machining (AFM). *Materials Today: Proceedings*, 56, 3108–3116.
19. Sharma, A. K., Venkatesh, G., Rajesha, S., & Kumar, P. (2015). Experimental investigations into ultrasonic-assisted abrasive flow machining (UAAFM) process. *The International Journal of Advanced Manufacturing Technology*, 80, 477–493.
20. Das, M., Jain, V. K., & Ghoshdastidar, P. S. (2011). The out-of-roundness of the internal surfaces of stainless-steel tubes finished by the rotational–magnetorheological abrasive flow finishing process. *Materials and Manufacturing Processes*, 26(8), 1073–1084.
21. Brar, B. S., Walia, R. S., & Singh, V. P. (2015). Electrochemical-aided abrasive flow machining (ECA 2 FM) process: a hybrid machining process. *The International Journal of Advanced Manufacturing Technology*, 79, 329–342.
22. Sankar, M. R., Mondal, S., Ramkumar, J., & Jain, V. K. (2009). Experimental investigations and modelling of drill bit-guided abrasive flow finishing (DBG-AFF) process. *The International Journal of Advanced Manufacturing Technology*, 42, 678–688.

10 Precision finishing innovations

Recent advances in abrasive flow machining, magnetic abrasive finishing, and magnetorheological finishing

Anant Bhardwaj, Parvesh Ali, Mohammed Omar Abdullah Ba Matraf, Abdul Wahab Hashmi, Shadab Ahmad, Yebing Tian, and Faiz Iqbal

10.1 INTRODUCTION

In the continually evolving realm of manufacturing and precision engineering, achieving flawless surface finishes has taken center stage. With industries seeking components that deliver superior performance and adhere to stringent tolerances, conventional finishing methods often prove inadequate in meeting these exceedingly high standards. As a result, innovative and advanced finishing processes have come to the forefront, pushing the limits of attainable surface quality. Among these state-of-the-art techniques, abrasive flow machining (AFM), magnetic abrasive finishing (MAF), and magnetorheological finishing (MRF) have emerged as pioneering solutions poised to transform the industry. These processes represent a new paradigm in surface finishing, harnessing the power of abrasives and magnetic forces to polish and refine workpieces with unparalleled precision.

AFM is a non-traditional finishing process in which a mixture of an abrasive polymer and gel is extruded due to the pressure difference within the internal surface of the workpiece, which results in effective machining [1, 2]. The process finds its application in many areas including in the aerodynamics, aerospace, biomedical, cosmetics, and manufacturing industries; as a result, the development in this field is rapid. With the development of new hybrids, the finishing is achieved up to the nano level. The major limitation of the AFM process is the low magnetorheological (MR) and large machining time (comparing traditional machining) for which various hybrids of AFM are developed [3, 4]. The next abrasive-based finishing process is MAF,

DOI: 10.1201/9781003496298-10

which is the prominent process of surface finishing where micro- and nano-finishing is achieved. The process uses a rotating magnetic brush made of carbonyl particles and abrasive particles placed along the workpiece surface which is to be finished. The abrasion mechanism of the abrasive and carbonyl particles results in the material removal process [5]. With the rapid growth of population, the demand for the continuous development of non-traditional machining has significantly increased. A significant effort has been made in improving the efficiency of finishing processes by using different abrasive types like alumina, Silicon Carbide, diamond calcite (calcium carbonate), emery (impure corundum), diamond dust (synthetic diamonds are used extensively), novaculite, pumice, iron (III) oxide, sand and corundum, which are used with different mesh sizes. Many researchers have used abrasives of different shapes to achieve an efficient finishing process [6–9].

Against this specific historical backdrop, the concept of MRF utilizing MR fluids has emerged. The inception of MRF dates back to the 1950s when Rabinow et al. first introduced the concept. Notably, a significant development occurred in 1995 when Kordonsky, W. I., et al., secured a patent (US patent) titled "Magnetorheological polishing devices and methods" [10], which marked a pivotal point in its advancement. Over the ensuing two decades, MRF gained traction, undergoing refinement and widespread adoption. In comparison to conventional mechanical finishing methods, MRF offers distinct advantages attributed to its adaptable interaction dynamics between tools and the workpiece. This characteristic engenders a ductile material removal process, enhancing precision and surface quality [11–13]. Abrasive-based finishing was probably used in the finishing of 3D printed parts, selective laser melting (SLM) projects, additive manufacturing products, and machining of industrial-based alloys [14, 15]. MAF and MRF have emerged as a modern tool in the finishing process. These two methods use less force than AFM, and thus have lesser material removal rate (MRR) but give a better micro/nano finish comparatively because of the gentle magnetic abrasive brush. Many researchers developed sintered media, plasma atomized particles, viscoelastic media, and magnetorheological shear-thickening polishing (MSRTP) media for increasing the efficiency of the process and searched for new avenues for finishing delicate, hard, brittle materials and the complex intricate shapes of components [16–20].

This review delves into the realm of abrasive-based advanced finishing processes, exploring the principles, capabilities, and emerging trends within AFM, MAF, and MRF. By examining these techniques, we aim to shed light on how researchers are dealing with precision manufacturing and providing engineers and manufacturers with novel tools to achieve levels of surface perfection once thought unattainable. As we embark on this exploration of abrasive-based advanced finishing processes, it is essential to understand their underlying mechanisms, applications, and the technological advancements that are propelling them forward. Through this examination, we will uncover the transformative potential of these techniques and gain insights into how they are driving innovation across industries.

10.2 PROCESS HYBRIDIZATIONS AND CLASSIFICATION

The classification of abrasive-based finishing processes targeted in this chapter may be studied from Figure 10.1. Broadly, abrasive-based finishing is divided into three

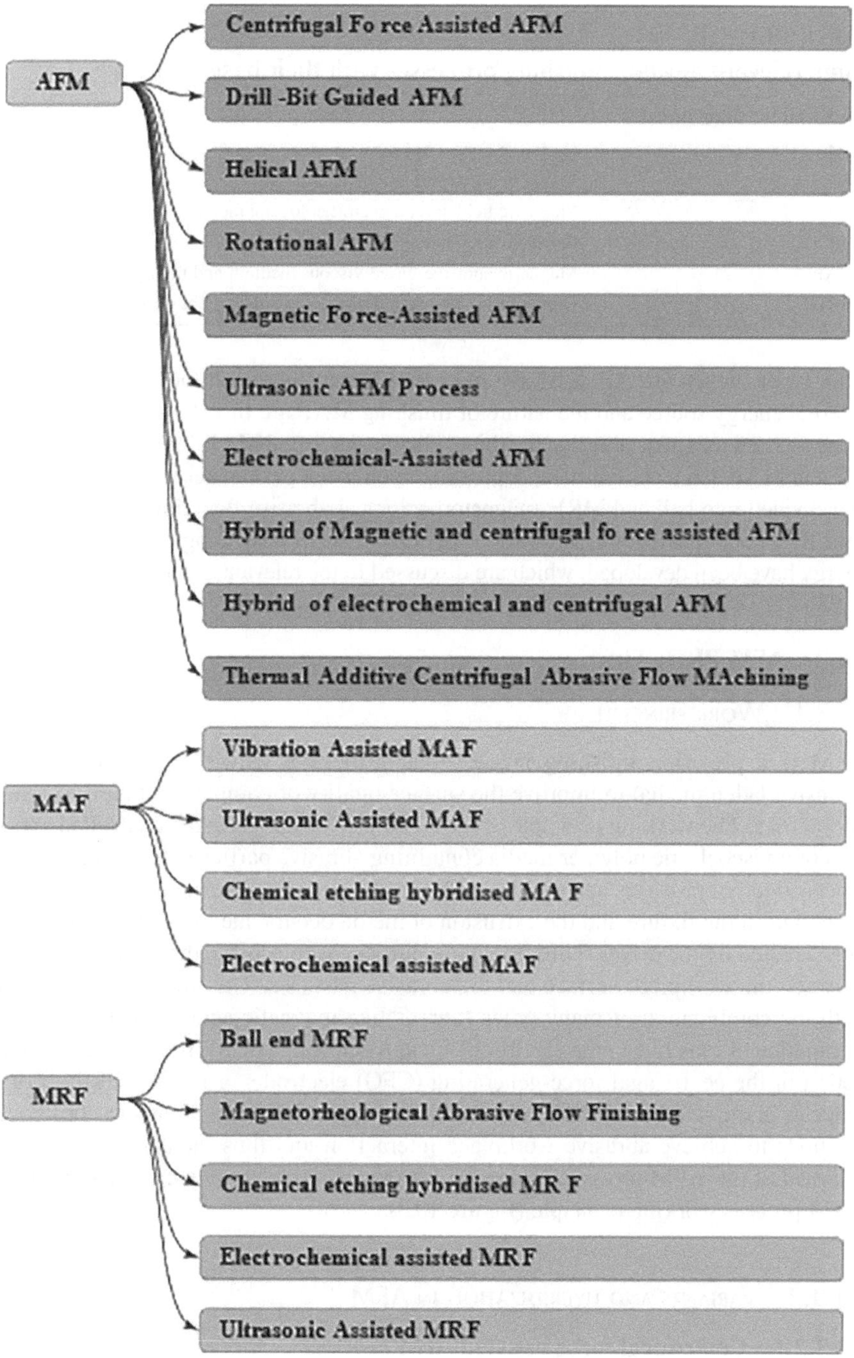

FIGURE 10.1 Various variant of AFM, MAF, MRF processes.

TABLE 10.1
Some relevant advance finishing processes with their basic principle

Advanced finishing process	Basic principle
AFM	High pressure and abrasive media
MAF	Magnetic force and magnetic abrasive
MRF	Magnetic field and magnetorheological media
MFP	Ferro-hydrodynamic behavior of MF
MRAFM	Magnetic abrasive-based viscous medium and high-pressure finishing
MSRTP	Carbonyl-based viscoelastic media polishing basically MRSTP media

main types, which are AFM, MAF, and MRF. Depending upon the requirement, the external energy source and the nature of finishing MAF are further divided into ultrasonic assisted magnetic abrasive finishing, chemical assisted MAF, and electrochemical assisted MAF. On a similar note, depending upon the experimental setup, MRF is further divided into ball end MRF, magnetorheological abrasive flow finishing, and magnetic float polishing (MFP) [21]. Various hybrids of AFM depending upon the source of energy have been developed, which are discussed in the relevant sections (Table 10.1).

10.3 AFM PROCESS

10.3.1 Work principle

AFM is a precision finishing process that utilizes a semi-solid abrasive media (abrasive-laden media) to improve the surface quality of complex and hard-to-reach geometries. The working principle of the AFM process involves the controlled extrusion of a viscoelastic polymer media containing abrasive particles through the workpiece's internal passages and surfaces. The workpiece whose surface is to be finished is set within the fixture and the extrusion of media occurs due to the pressure difference created by the driver. Thus, the main purpose is to achieve the abrasion process between the workpiece surface and the abrasive particles. This abrasion is achieved with the combination of many other sources like magnetic action, electrochemical dissipation of ions like in the electrochemical machining (ECM) process, centrifugal action of the centrifugal force-generating (CFG) electrode, or rotation of the workpiece as in the rotational AFM process. The passage of time has shown various novel methods to achieve abrasive workpiece interaction and thus increase the material removal of the AFM process. Figure 10.2 provides a step-by-step explanation of the AFM process working principle (Figure 10.3).

10.3.2 Variants and hybridization in AFM

10.3.2.1 Centrifugal force assisted AFM process

The centrifugal force assisted abrasive flow machining (CFAAFM) process [22–24] includes a CFG rod aligned along the axis of the cylinder of the process. When the rod rotates, it gives a centrifugal effect to the abrasive present in the media.

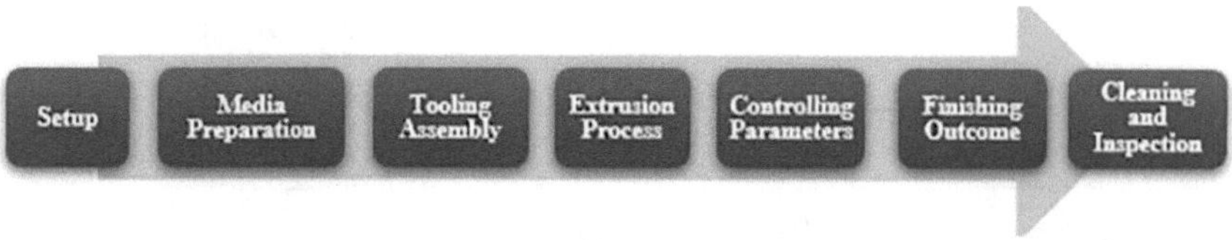

FIGURE 10.2 Step-by-step explanation of the AFM process.

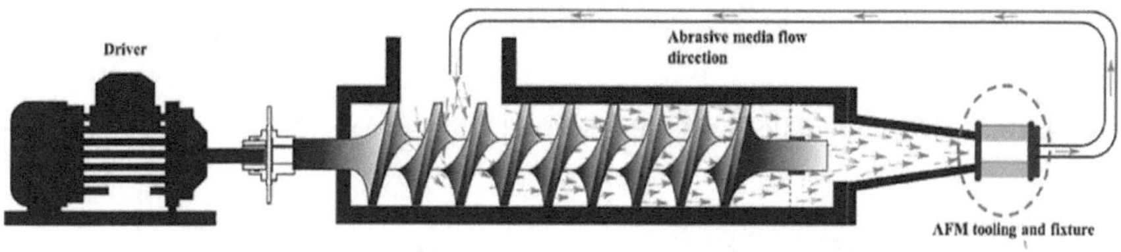

FIGURE 10.3 The schematic of a one-way AFM process [24].

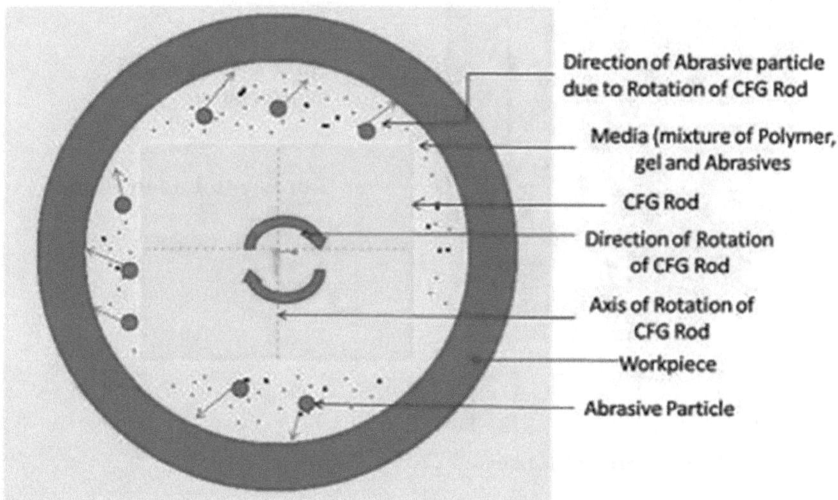

FIGURE 10.4 The CFAAFM process [24].

This abrasive hits the workpiece's surface, resulting in practical and desired machining. The resultant combination of the reciprocation and the centrifugal force results in effective machining, thus improving the MR and ΔRa. Figure 10.4 shows the principle of CFAAFM.

10.3.2.2 Drill-bit-guided AFM process

The next variant in the AFM field is the drill-bit-guided abrasive flow machining process (DBGAFM) [25]. This process is similar to CFAAFM. Still, the difference is that instead of a rotatory centrifugal rod, a fixed drill bit aligns coaxially with the

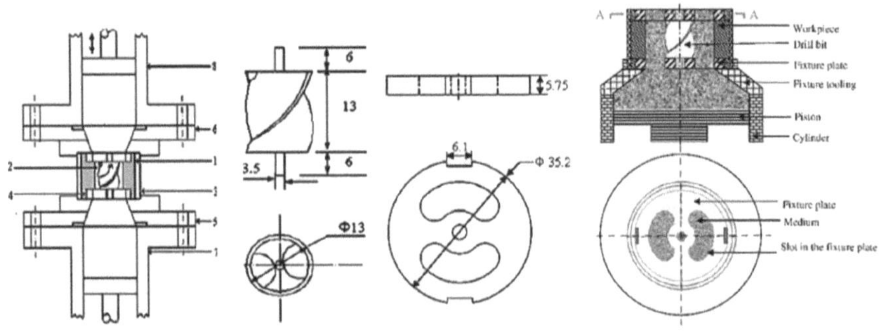

FIGURE 10.5 Drill-bit-guided abrasive flow machining process [25].

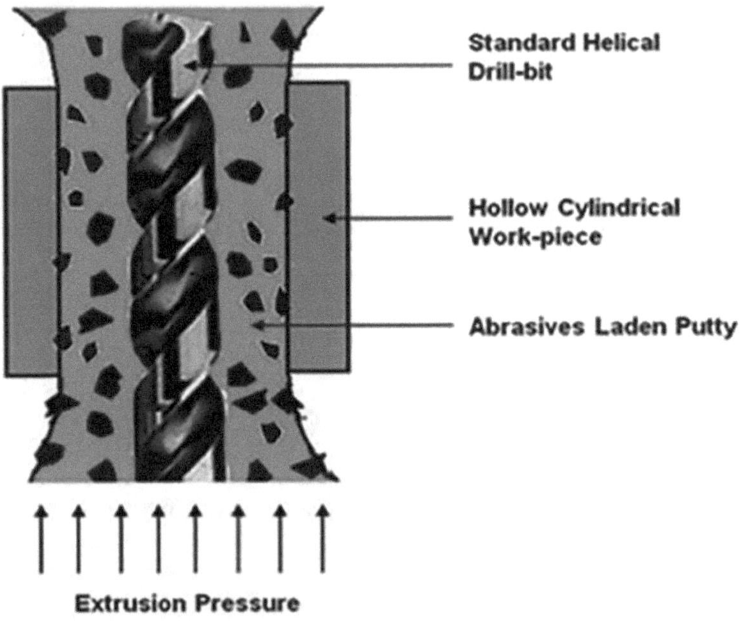

FIGURE 10.6 Helical abrasive flow machining.

cylinder, which provides a restricted passage to the abrasive, which results in the MR and the effective machining process. Figure 10.5 shows DBGAFM.

10.3.2.3 Helical AFM process

In the helical AFM [26], a drill bit is rotated to provide the centrifugal action and machining in the restrictive path due to the helical profile with the advantage of the conventional AFM process. Figure 10.6 shows the helical AFM process.

10.3.2.4 Rotational AFM process

The workpiece's course provides the effective rotation between the workpiece and the abrasive, which results in the same effect that can be achieved by giving the

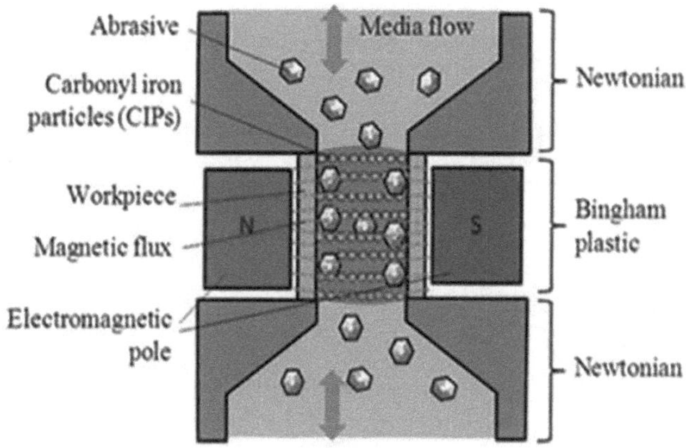

FIGURE 10.7 Magnetic force assisted abrasive flow machining.

CFAAFM. So the workpiece is rotated by a specific mechanism. Thus, effective machining occurs by combining the media's reciprocation and the rotation of the workpiece [27].

10.3.2.5 Magnetic force assisted AFM process

The first hybridization in the process is the magnetic force assisted AFM process [28]. This process combines a magnetic force and the conventional machining process. The process involves including the carbonyl particles in the media affected by the magnetic force. The carbonyl particles hit the abrasive, and this abrasive hits the surface which is to be finished, thus making the process effective. Figure 10.7 shows the principle of magnetic AFM.

10.3.2.6 Ultrasonic AFM process

The next hybrid is the ultrasonic AFM process [29]. Here, ultrasonic vibrations are given perpendicular to the axis of the media's reciprocation. These longitudinal vibrations and the conventional AFM process provide effective machining and the desired surface finish.

10.3.2.7 Electrochemical assisted AFM process

The hybridization is the electrochemical assisted AFM (ECMAAFM) [30]. This hybridization aims to dissolve the anodic workpiece by the electrolyte, thus making the requisite material removal. The workpiece is made of an anode, and the cylinder is made of a cathode. The cylinder is generally made of copper for this process. The major limitation of this process is the uncontrolled dissolution of the workpiece. So the selection of the electrolyte is of paramount importance. Figure 10.8 shows EFAAFM.

10.3.2.8 Various other hybrid AFM processes

Hybrid of MFAFM and CFAAFM: this process involves using a permanent magnet that attracts the laden media forced by the centrifugal force produced by the rotating

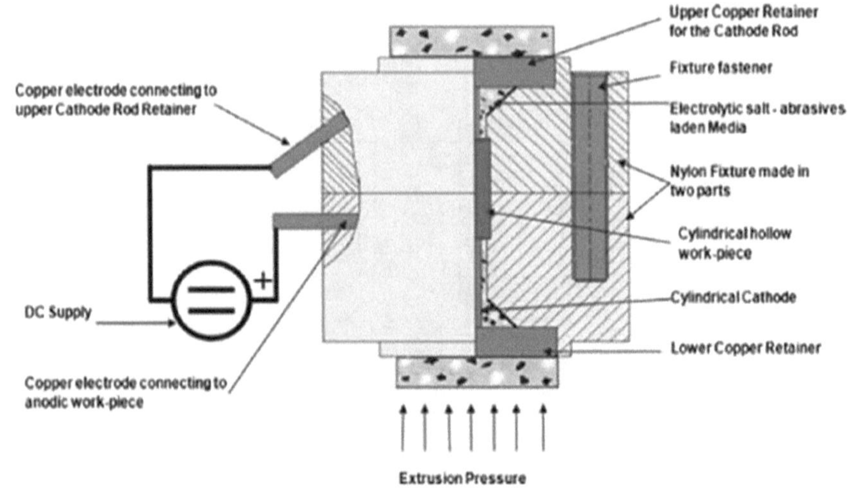

FIGURE 10.8 Electrochemical force assisted abrasive flow machining process.

rod placed coaxially inside the cylinder. This results in the effective machining of the process with CFAAFM, MFAFM, and the conventional AFM process. The next hybridization in the process is the hybrid electrochemical and CFAAFM [31].

Hybrid of ECMAFM and CFAAFM process: ECMAFM and CFAAFM and AFM have been clubbed together to rule over the low MR of the conventional AFM process [24]. Researchers have used an increased amount of KI in the media so that the media's anodic dissolving property increases; also, the rod responsible for the centrifugal action workpiece is made one of the cathodes. So the combination is helpful in this case.

Thermal additive centrifugal force assisted abrasive flow machining process (TACAFM): This brings the advantage of electric discharge machining and the *TACAFM* process takes this into account [32–34]. In this process, a non-conductive electrolyte is used. The spark is produced when the gap between the rod and the workpiece wall is tiny. The current is supplied in the centrifugal rod due to the pulse voltage. The potential difference is created between the rod and workpiece. The MR occurs due to the spark generated by the process. The TACAFM process needs EDM power supply and an EDM controller for the spark generation along with the CFAAFM [33].

10.4 MAGNETIC ABRASIVE FINISHING

10.4.1 Work Principle

MAF involves the nano-finishing of the work surfaces using carbonyl particles, abrasives, and a rotating magnet [36]. The carbonyl and the abrasive particles are sintered, mechanically alloyed, plasma atomized, or loose abrasives. Sometimes machining fluids are added for cooling and lubrication and for temporary binding when using loose magnetic abrasives [6, 37, 38]. A rotating electromagnetic field is applied to the magnetic abrasives, which results in the formation of an electromagnetic brush of

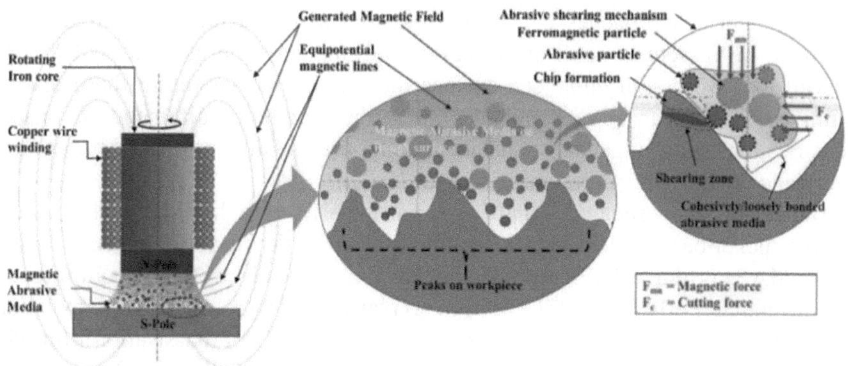

FIGURE 10.9 Schematic diagram of the MAF process [35].

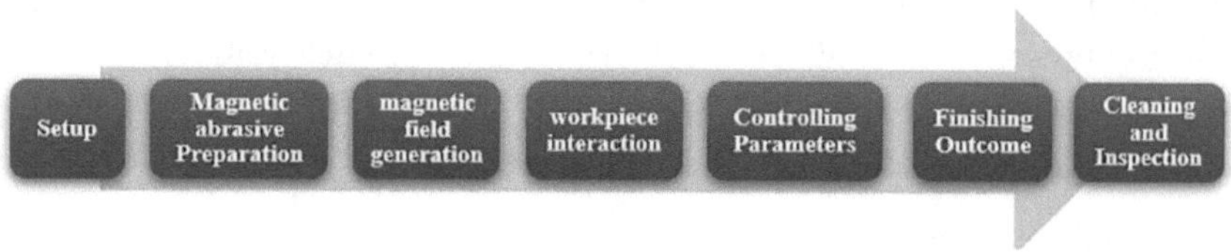

FIGURE 10.10 Schematic diagram of the MAF process.

carbonyl particle media [35, 39]. This brush drives the workpiece and results in the surface finishing process (Figures 10.9 and 10.10).

10.4.2 Variants and hybridization in MAF

MAF as shown in Figure 10.11 is a highly versatile precision machining process that offers various adaptations and hybridizations to meet specific surface finishing needs [40–42]. One variant is bonded MAF [43], where abrasive particles are affixed to a carrier medium, creating uniform abrasive action suitable for flat or contoured surfaces. In contrast, unbonded MAF suspends abrasive particles in a magnetic fluid (MF), allowing for intricate geometries and complex workpiece shapes [44]. The choice between permanent magnets and electromagnets for generating the magnetic field has distinct advantages. Permanent magnets offer simplicity, while electromagnets provide precise control over field strength and distribution.

Introducing mechanical vibrations in vibration assisted MAF enhances the material removal efficiency by dislodging debris from abrasive particles and the workpiece surface [45]. Ultrasonic assisted MAF combines ultrasonic vibrations with the magnetic field, further boosting MRRs (shown in Figure 10.12) and overall finishing performance [43]. MAF can be combined with other processes in a hybrid approach, including chemical etching and electrochemical assisted finishing. In chemical etching, abrasive particles are combined with chemical agents to enable controlled material removal, while electrochemical assisted MAF combines ECM with MAF to achieve controlled material removal and superior surface finish [40].

These hybrid approaches can be seamlessly integrated into both bonded and unbonded MAF setups, resulting in several advantages such as increased MRRs, reduced tool wear, and enhanced surface quality. By incorporating these diverse techniques, MAF can be customized to meet the specific requirements of various industries, ensuring impeccable surface quality, precise dimensional accuracy, and compliance with stringent finishing standards.

Addressing the efficiency challenge, particularly with high-hardness materials like nickel-based alloys, is the realm of electrolytic-magnetic abrasive finishing (E-MAF) [46]. E-MAF combines the advantages of electrolytic machining and MAF to enhance efficiency [47]. Electrolysis plays a critical role in E-MAF by inducing anodic dissolution on the workpiece's surface, forming a passivation film with lower hardness than the substrate. MAF subsequently removes this film, reducing the grinding difficulty. Liu et al. [46] introduced a tool for E-MAF, emphasizing the benefits of ECM based on anodic dissolution, such as high MRRs and minimal tool wear. However, ensuring that the electrolyte reaches the working area remains a challenge in E-MAF, necessitating careful consideration in tool design.

10.5 MAGNETORHEOLOGICAL FINISHING

10.5.1 Work principle

MRF is a precision finishing method that leverages a MR fluid to attain exceptional surface quality and precise geometric accuracy [11, 12]. This process hinges on the application of a magnetic field to manipulate the behavior of the MR fluid, which comprises a carrier fluid, magnetizable particles, and abrasive particles. When a magnetic field is introduced, the magnetizable particles within the MR fluid align themselves along the field lines, as illustrated in Figure 10.15(c), leading to a significant increase in the fluid's viscosity. This phenomenon is known as the Bingham effect, and transforms the MR fluid into a semi-solid state, enabling it to behave as a precise and controlled abrasive media [48–50]. During MRF, the workpiece is brought into contact with the MR fluid. As the workpiece and fluid interact, relative motion ensues. The stiffened MR fluid adapts to the shape of the workpiece, allowing abrasive particles to gently remove material through a combination of mechanical abrasion and, potentially, chemical interactions [51]. The process parameters, including magnetic field strength, rotational speed, and finishing time, are carefully adjusted to achieve the desired MRR and surface roughness.

MRF excels in finishing complex, curved, and high-value surfaces with sub-micron precision. It is widely used in applications like optics, semiconductors, and aerospace components, where surface quality and dimensional accuracy are critical. MRF's unique ability to adapt to intricate geometries and the controlled removal of material make it a powerful tool for achieving exceptional surface finishes and meeting stringent industry requirements.

10.5.2 Variants and hybridization in MRF

MRF, a cutting-edge precision machining process, exhibits remarkable versatility through a variety of variants and hybridizations that harness its unique properties

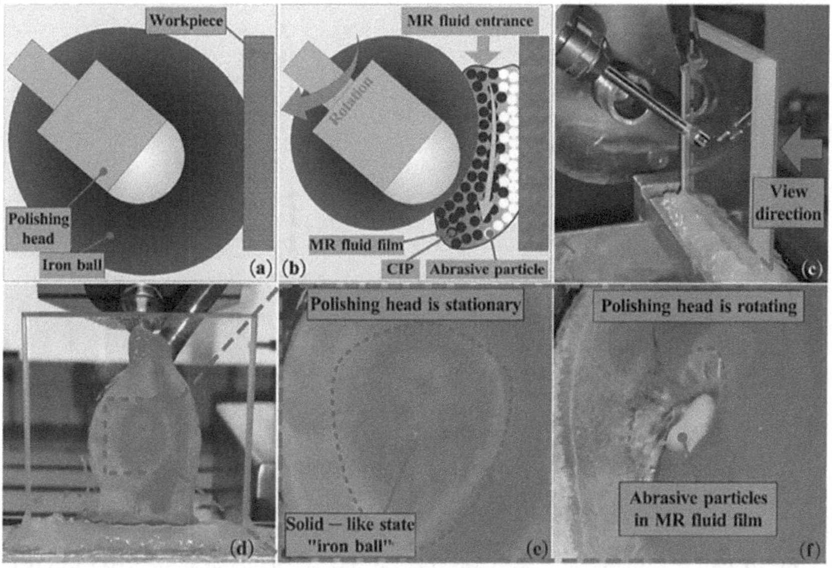

FIGURE 10.11 Schematic diagram of material removal mechanism. (a) Polishing zone when the polishing head is stationary; (b) polishing zone when the polishing head is rotating; (c) view direction; (d) back of workpiece surface; (e) iron ball with solid-like state; (f) MR fluid film [52].

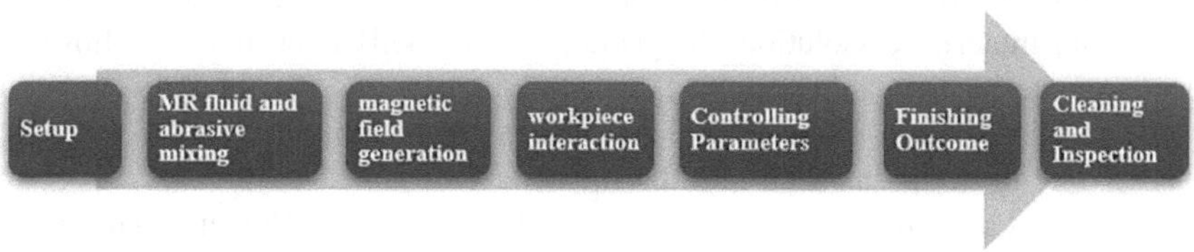

FIGURE 10.12 Schematic diagram of the MAF process.

to achieve specialized surface finishing outcomes. These innovative approaches broaden MRF's scope and enhance its effectiveness in meeting specific requirements.

One variant, "Ball End MRF," employs a ball-ended tool interacting with the MR fluid and workpiece under a magnetic field [53]. This configuration enables efficient and controlled material removal, proving especially advantageous on curved or intricate surfaces. The ball end ensures uniform abrasion, making it ideal for achieving consistent finishes on complex geometries. "Magnetorheological Abrasive Flow Machining (MRAFM)" seamlessly merges MRF principles with AFM [54, 55]. The MR fluid, combined with abrasive particles, is propelled through intricate internal passages of a workpiece. The magnetic field governs the abrasive-laden MR fluid, resulting in precise material removal within convoluted channels and passages [19]. The integration of "Chemical Etching in MRF" takes a hybrid approach by incorporating chemical agents into the MR fluid [56]. The magnetic field disperses the chemical agents uniformly, enhancing material removal through both abrasive and chemical actions. This synergistic combination enables controlled material removal

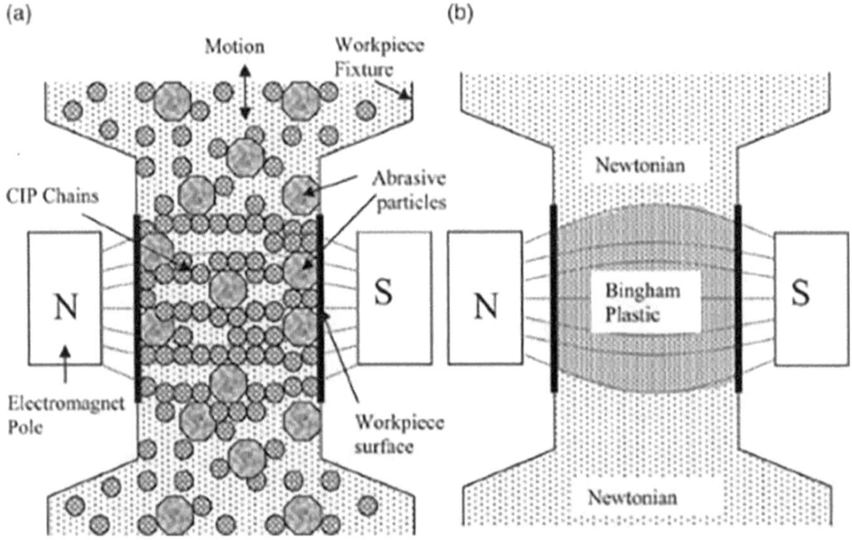

FIGURE 10.13 Schematic of magnetorheological abrasive flow finishing (MRAFF) [54].

and heightened surface quality, which is particularly advantageous for materials susceptible to chemical reactions. Furthermore, mixing electrochemical in MRF, ties ECM with MRF [57]. The magnetic field influences electrolyte distribution, augmenting material dissolution alongside abrasion. This hybridization allows for selective material removal, improved surface integrity, and the flexibility inherent to MRF. "Ultrasonic assisted MRF" introduces ultrasonic vibrations to the MRF process. These vibrations facilitate the movement of abrasive particles across the workpiece, reducing friction and enhancing material removal efficiency. The combination of ultrasonic vibrations and the magnetic field enhances process effectiveness, leading to superior surface finishes and reduced machining times.

These variants and hybridizations empower MRF to cater to an array of specific applications. By integrating MRF with other processes, engineers and manufacturers can attain intricate finishes, selective material removal, and superior surface quality across diverse workpiece geometries and materials. These adaptations underscore MRF's adaptability and potential to address evolving industry demands for precision, efficiency, and quality in surface finishing, establishing it as a pivotal player in modern machining practices. Figure 10.13 shows hybrids of the MRF and AFM process, which is magnetorheological AFM.

10.5.3 Abrasive Media and Its Preparation Techniques

Abrasives play a pivotal role in surface finishing, especially when they are carried and supported by a media particle. Media itself is instrumental in facilitating abrasive action and transporting the removed material, thereby enhancing the abrasion process. Generally, media in abrasive-based finishing serve two fundamental purposes. First, they act as a holder for abrasive particles, enabling them to exert

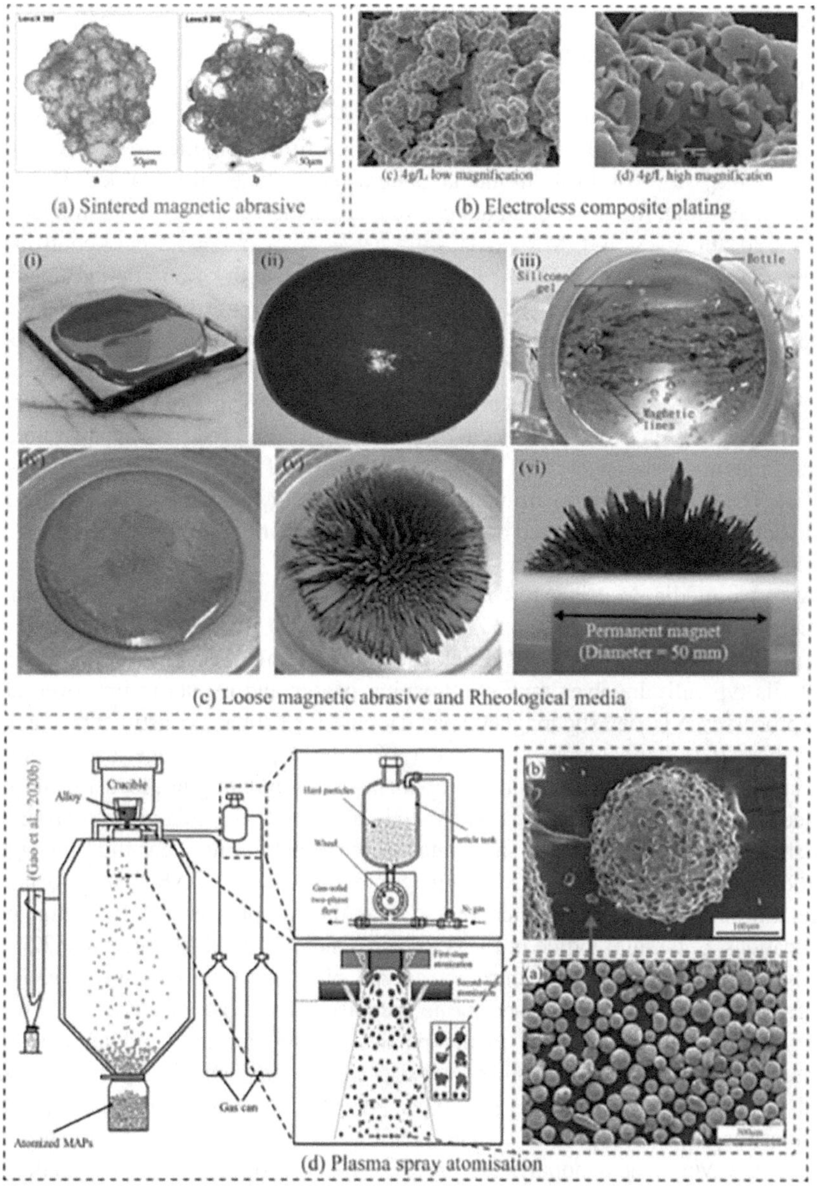

FIGURE 10.14 (a) Sintered [59]; (b) electroless composite [60]; (c) (i) by Ahmad et al. [35]; (ii) by Li Wenhui et al. [61]; (iii) magnetic lines of steel grits in the silicone gel [62]; (iv) condition of MR fluid without magnetic field; (v) with magnetic field and (vi) side view [41].

abrasion on the workpiece surface. Second, they function as a medium for transporting the removed material. Over time, the development of media has progressed rapidly, with new properties emerging based on chemical composition and the types of abrasives used. In many cases, media is a blend of polymer gel and abrasives in AFM processes [4, 19]. In contrast, when employed in MAF processes, it typically consists of carbonyl particles, which influence the media's magnetic characteristics, thereby accelerating the finishing processes [6, 36, 58].

Rheology is the study of how materials deform under mechanical loading, encompassing aspects like yield stress, relaxation time, compliance, and viscosity. Viscosity plays a pivotal role in media as abrasive particles are retained. However, during machining, the heat generated can reduce media viscosity, impacting its ability to hold abrasives [63]. MR fluids are a category of magnetic field-controllable fluids whose rheological behavior depends on the strength of an applied magnetic field. Rheological properties, such as viscosity and shear stress, typically vary with factors such as chemical composition and temperature [64, 65]. Figure 10.14 presents various abrasives used in MAF and MRF. Plasma spraying involves subjecting gases like argon or hydrogen to a powerful electric arc, resulting in the generation of high-temperature plasma. This intense heat energy is directed onto a substrate containing magnetic abrasives, leading to the formation of magnetic abrasive particles [8, 66]. These developments in media and their rheological properties significantly contribute to more effective and controlled surface finishing processes, especially in applications like AFM and MAF.

Abrasive media used in AFM is often a semi-solid mixture comprising abrasive particles and a viscoelastic gel. Figure 10.15(a) shows making of CNT-based and simple abrasive media for AFM developed by Ali et al. [32]. The use of this type of media typically involves multiple finishing passes and results in improved abrasion processes. Viscoelastic magnetic particles are the primary constituents of a magnetic slurry [67]. For example, Dixit et al. (2020) developed a hydrogel-based abrasive media for AFM of polymers using components like xanthan gum, locust bean gum, fumed silica, water, and abrasives. Their study revealed that media viscosity is high at low shear rates but decreases as shear rates increase, exhibiting shear-thinning behavior [68], as shown in Figure 10.15(b). Slurries are also utilized in AFM, with variations depending on the specific application, such as the use of iron powder in magnetic applications, potassium iodide in ECM, or kerosene in TACAFM processes [32].

Shear-thickening abrasive slurry is a type of dilatant fluid where the viscosity generally increases as the shear rate increases. This viscosity rise leads to the cross-linking and bonding of media particles with carbonyl and abrasive particles. Typically, this type of slurry consists of carbonyl and SiC particles and is used in shear-thickening MAF processes [49, 69, 70]. The system's composition includes magnetic nanoparticles, a carrying liquid, and surfactants, as illustrated in Figure 10.2. Magnetic nanoparticles serve as the source of magnetism within the MF. Surfactants are adsorbed onto the surfaces of these magnetic nanoparticles to prevent their aggregation in a free state. These surfactant-coated magnetic nanoparticles are suspended within the carrying fluid, giving the MF its fluidity.

As depicted in Figure 10.15(c), when the applied magnetic field strength remains below a certain critical value ($H_{c,n}$), the magnetic nanoparticles in the MF remain evenly dispersed within the base fluid. However, when the magnetic field strength surpasses $H_{c,n}$, the nanoparticles within the MF tend to aggregate or align themselves into chains along the magnetic field direction. Eventually, the relative magnetic permeability (RI) saturates at a specific magnetic field strength, and further increases in

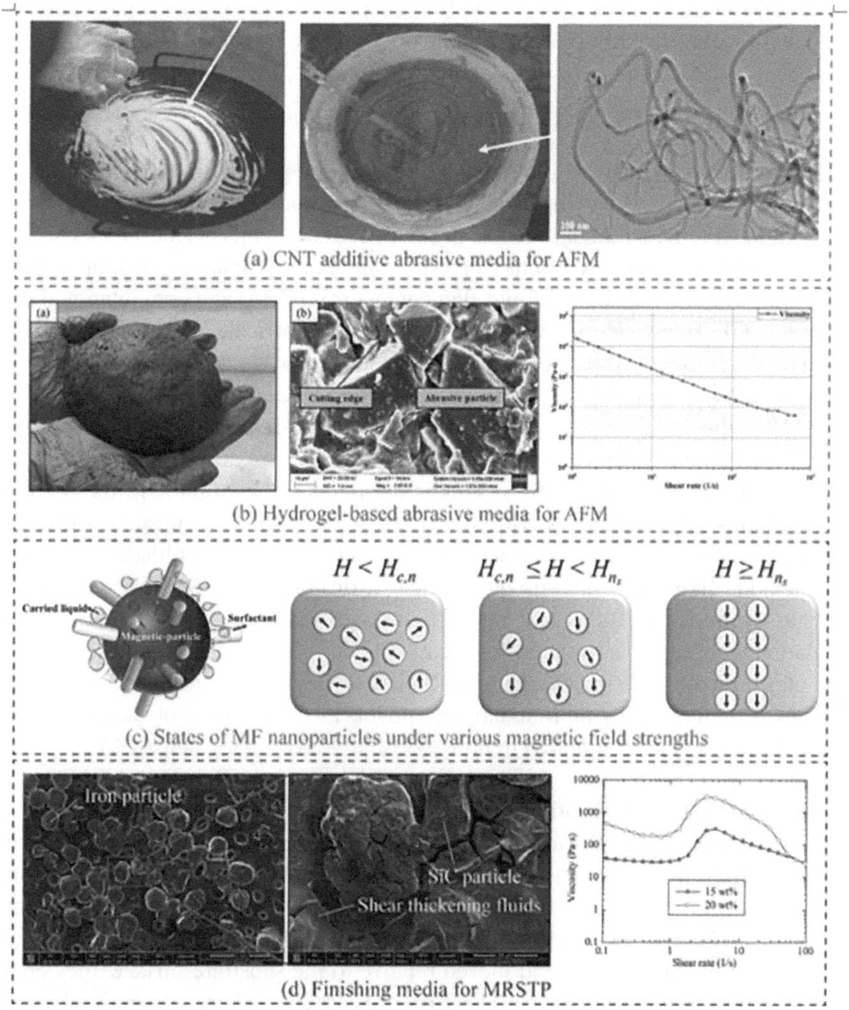

FIGURE 10.15 (a) CNT-based media for AFM; (b) hydrogel-based media for AFM; (c) states of MF nanoparticles under various magnetic field strengths; (d) rheological curves of two solids contains SEM images of finishing media [69].

magnetic force do not impact it [71]. In a study conducted by Fan et al. in 2020, they prepared a finishing media by combining shear-thickening fluids (STFs), carbonyl iron particles, and SiC particles. They used polyethylene glycol (PEG) with a molar mass of 200 g/mol as the base fluid. Fumed silica, with primary particle sizes of 50 nm, was directly dispersed into PEG 200. This mixture underwent mechanical blending for 60 minutes at 3000 rpm and a temperature of 80°C, controlled by an oil bath. Rheological tests were conducted on the STFs at concentrations of 15 wt% and 20 wt% using a rotational rheometer (MCR 302, Germany) at 25°C. Figure 10.15(d) illustrates the relationship between viscosity and shear rate. It's worth noting that higher concentration STFs exhibited both higher initial viscosity and maximum viscosity, with values reaching 310.6 Pa s and 3070.6 Pa s, respectively [69].

10.6 SIMULATION AND MODELING

Mathematical and numerical modeling techniques are pivotal in understanding and enhancing the performance of advanced finishing processes like MAF, MRF, and AFM. In MAF, mathematical models predict MRRs by considering factors such as abrasive particle properties, contact pressure, and sliding distance. Simultaneously, numerical simulations, notably finite element analysis (FEA) and computational fluid dynamics (CFD), provide in-depth insights into the intricate interactions between magnetic fields, abrasive particles, and workpiece materials. These simulations help optimize process parameters to achieve desired MRRs and surface finishes.

In MRF and MAF, mathematical models account for parameters like magnetic field strength and abrasive concentration, while numerical simulations examine abrasive particle behavior within magnetic abrasives or MR fluids. In the realm of advanced finishing methods, MRR models which take some factors such as viscosity and flow rates into account are essential. Researchers employ sophisticated simulation methods like FEA and smooth particle hydrodynamics (SPH) to capture the intricate dynamics of fluids and interactions with materials. These modeling and simulation techniques are crucial for creating accurate predictive models, which, in turn, play a pivotal role in enhancing MRR and achieving better surface roughness.

Many researchers rely on foundational models as the building blocks for predicting MRR and surface roughness in processes like MAF, MRF, and AFM.

i. Classical Preston equation (for MRR):

The classical Preston equation, which is commonly employed for describing the MRR in abrasive-based processes, posits that MRR has a linear relationship with both polishing pressure (*P*) and relative velocity (*V*), as depicted in Eq. (10.1). This relationship stems from the notion that when abrasives are subjected to a polishing pressure and moved relative to the substrate surface, they create plow grooves on the surface. These grooves are formed because of the abrasives being pressed into the substrate to a specific depth, and they are essential for facilitating the continuous polishing process, where a multitude of abrasives collectively contribute to achieving the final material removal [72, 73].

$$\boldsymbol{MRR} = \boldsymbol{K} \times \boldsymbol{P} \times \boldsymbol{V} \tag{10.1}$$

where MRR is the material removal rate, $\boldsymbol{K}$ is the Preston's constant, $\boldsymbol{P}$ is the contact pressure.

$\boldsymbol{V}$ is the sliding relative velocity.

ii. Elasto-plastic deformation model (MRR):

In processes like MAF, understanding the workpiece's deformation behavior under applied forces is crucial. Elasto-plastic deformation models, like the Johnson-Cook model, predict how materials deform and eventually fail under loading conditions, contributing to MRR modeling [74, 75].

iii. Empirical roughness models (surface roughness):

Empirical models, such as the empirical Ra model, are often used to predict surface roughness. They relate surface roughness (Ra) to factors like abrasive particle size, concentration, and process parameters. These models can be adapted for surface roughness predictions in various finishing processes. For example:

$$\textbf{Surface Roughness} = f\begin{pmatrix}\textbf{Abrasive, charachteristics, concentration,}\\ \textbf{process parameters}\end{pmatrix} \quad (10.2)$$

iv. Fluid flow models (for surface roughness):

For processes like AFM, modeling the behavior of the abrasive medium is fundamental. Fluid flow models, including the Navier-Stokes equations, describe the movement of the abrasive medium within the workpiece. By considering factors like viscosity and flow rate, these models can predict surface roughness. See various viscosity models in Table 10.2. The mathematical modeling and computation in case of AFM has been done based on pressure variation and Coriolis effect on the workpiece produced due to simultaneous rotation and extrusion of the media within the workpiece [76, 77].

v. Magnetic field models (MAF and MRF):

In MAF and MRF, understanding the magnetic field's impact on abrasive particles is essential. Magnetic field models describe the distribution and strength of the magnetic field within the processing area. These models can be used to predict how magnetic forces influence abrasive particle movement and, consequently, material removal and surface quality. The intensity of the magnetic field H is a gradient of the magnetic scalar potential φ and is expressed in two subsequent equations as:

$$H = -\nabla\varphi \quad (10.3)$$

Based on the above assumptions, in the axisymmetric form, the governing equation of the process becomes:

$$\frac{1}{r}\frac{\partial}{\partial r}\left[\tau\mu_r\frac{\partial\phi}{\partial r}\right]+\frac{\partial}{\partial z}\left[\mu_r\frac{\partial\phi}{\partial r}\right] \quad (10.4)$$

where φ is a magnetic scalar potential and μ_r is the relative permeability of Magnetic abrasive particles (MAPs). The permeability of most ferromagnetic materials is not constant and varies with the magnetic field. It implies that Eq. 10.4 is non-linear in nature.

TABLE 10.2
Viscosity model

Models	Viscosity relation	Remarks	Reference
Bingham model	$\tau = \tau_0 + \mu_p \gamma$	where τ is the shear stress, τ_0 is initial shear stress, μ_p is the curve slope named plastic viscosity, while γ is the shear rate	[50]
Biviscosity model	$\begin{cases} \tau = \eta_1 \dfrac{du}{dy}, \lvert\tau\rvert < \tau_y \\ \tau = \eta_1 \dfrac{du}{dy} + \tau_l sgn\left(\dfrac{du}{dy}\right), \lvert\tau\rvert > \tau_y \end{cases}$	where τ_y is the apparent yield stress, $\tau_1 = \left(\tau_y\left(1 - \eta_2 / \eta_1\right)\right)$ is the intercept stress, η_1 and η_2 are the viscosity coefficients of the fluid exposed to a stress that is either less than or greater than the apparent yield stress	[50]
Power law	$\eta = k\dot{\gamma}^{c-1}$	where k represents the flow consistency, c is the flow behavior index when $c < 1$, the fluid is shear-thinning, and the apparent viscosity drops as the shear rate increases when $c < 1$, the fluid is shear-thickening, and the apparent viscosity increases with shear rate: when $c = 1$, the fluid exhibits typical Newtonian behavior	[78, 79]
Casson model	$\eta^{\frac{1}{2}} = \eta_\infty^{\frac{1}{2}} + \tau_c^{\frac{1}{2}} \dot{\gamma}^{-\frac{1}{2}}$	where, η_∞ is the Casson viscosity, η_c the Casson yield stress	[80, 81]
Herschel-Bulkley model	$\eta = \begin{cases} \eta_0, \lvert\tau\rvert < \tau_y \\ \tau = \dfrac{1}{\dot{\gamma}}\left(\tau_0 + k\left(\dot{\gamma}^c - \left(\dfrac{\tau_0}{\eta_0}\right)^c\right)\right), \lvert\tau\rvert \geq \tau_y \end{cases}$	where τ_y represents the yield stress, τ_0 is a relatively large initial viscosity to evaluate the "rigid-like" of MR fluids in a low strain rate regime $\dot{\gamma} \leq \dfrac{\tau_0}{\eta_0}$ when the instantaneous stress is less than yield stress, namely $\tau < \tau_y$. When the critical yield stress is achieved, the viscosity transforms to the power law; k and c represent the flow consistency and behavior index, respectively. When $c = 1$, the model is degenerated into the Bingham model	[82, 83]

10.7 APPLICATION AND LIMITATION OF PROCESSES

The abrasive-based finishing finds its application in major domains of manufacturing. The process of AFM is used for finishing of intrinsic and hard-to-reach surfaces while MAF is used in the nano-finishing of the surfaces [84], [85]. The process finds its major application in the biomedical, aerospace, cosmetics, and ancillary industries. The AFM finds its core application in the finishing of microchannels (MEMs) and the finishing of confocal channels formed by the EDM, LBM, and EBM processes [24, 33, 55, 86]. Abrasive-based finishing is majorly finding applications in the finishing of additively manufactured products made by SLM and 3D printing. The advantage of MAF is that it can be applicable on any surface as it does not require the workpiece to be conductive, which are major limitations of TACAFM processes [24]. MAF plays a prominent role in the finishing of semiconductor devices, defense equipment, and various products in the cosmetic industry where surface finishing is required. Figure 10.16 shows various applications of abrasive-based finishing processes.

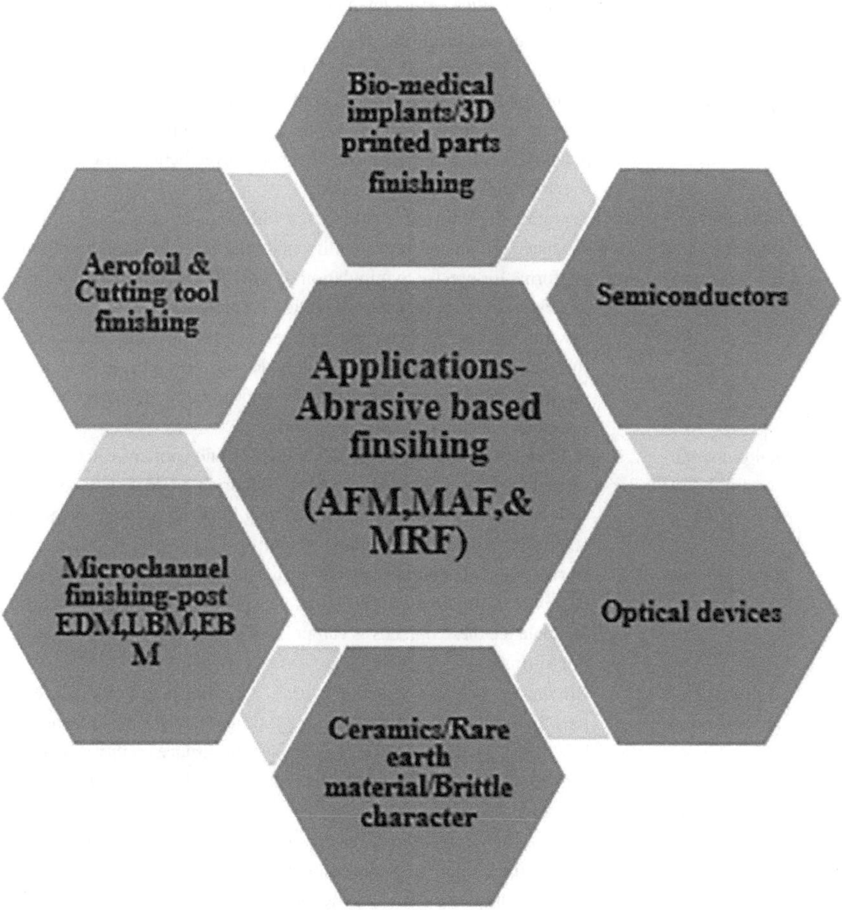

FIGURE 10.16 Various applications of abrasive-based finishing processes.

10.8 FUTURE PERSPECTIVE AND AVENUES

Looking ahead, these finishing techniques—AFM, MAF, and MRF—hold promising avenues for further advancement and application in modern manufacturing. These techniques hold immense potential to revolutionize the manufacturing landscape by offering tailored surface finishes, precision, and efficiency, aligning with the demands of modern industries and the pursuit of excellence in surface quality.

Table 10.3 provides a few future scope that can be considered.

TABLE 10.3
Future scopes

S. No	Scopes	Description
1	Advanced materials	As industries continue to adopt new materials, the adaptability of these techniques will be crucial. Exploring their effectiveness on advanced materials like composites, ceramics, and high-strength alloys can open up new possibilities for achieving high-quality surface finishes
2	Nanofinishing	Leveraging nanotechnology can lead to higher demand of finer surface finishes with improved accuracy and precision
3	Hybrid approaches	Combining finishing techniques with other processes, such as laser machining, electrochemical polishing, or ultrasonic vibrations, can potentially result in synergistic effects and even better surface quality, overcoming individual limitations
4	Automated and robotic systems	Integrating these finishing techniques with automation and robotics can enhance consistency, repeatability, and efficiency. Developing robotic platforms for precise manipulation of workpieces can streamline the processes and enable mass production scenarios
5	In situ monitoring and control	Real-time monitoring and control systems can ensure that the desired surface finish is achieved consistently. Research into integrating sensors, machine vision, and adaptive control algorithms can enhance the quality assurance aspect of these techniques
6	Environmental sustainability	Focus on environmentally friendly abrasives and coolants can contribute to sustainable manufacturing practices. Exploring biodegradable or reusable abrasive media and eco-friendly finishing fluids aligns with the growing emphasis on environmental responsibility
7	Multi-objective optimization	Consideration of multiple objectives such as surface finish, MRR, and energy consumption will be pivotal. Developing optimization algorithms that balance these objectives can provide a holistic approach to process improvement
8	3D printing applications	With the rise of additive manufacturing, these finishing techniques can play a vital role in post-processing 3D printed parts. Research can explore how AFM, MAF, and MRF can enhance the surface quality, accuracy, and functional properties of additively manufactured components
9	Micro and nano components	Exploring the scalability of these techniques to micro and nano components is essential. Industries like electronics and medical devices require ultra-small, high-precision parts, making it imperative to adapt these techniques for such applications

10.9 CONCLUSION

This review underscores the significance and potential of AFM, MAF, and MRF within contemporary manufacturing. AFM finds diverse applications in automotive, industrial, aerospace, and biomedical sectors, particularly for intricate profiles on both internal and external surfaces. MAF introduces superior surface quality for nonferrous materials like brass, though its efficiency decreases with mild steel and iron-containing materials. The hybrid AFM approach emerges as a solution to traditional AFM limitations, enhancing performance without compromising on material removal parameters. Despite advancements, room for AFM development persists, focused on achieving enhanced polish and cost-effectiveness. MAF parameters, notably working gap and rotational speed, influence MRRs and surface roughness, especially for non-magnetic stainless steel. In MRF, research navigates challenges such as magnetic particle dispersion and settling. Optimization of MR fluids is pursued for efficacy across materials. A deeper understanding of material removal mechanisms and numerical models is vital for MRF evolution. Composite MRF techniques promise elevated finishing quality, demanding exploration for scalability.

These methods, AFM, MAF, and MRF, tackle challenges across industries, contributing to precision, efficiency, and adaptability in modern manufacturing. Research and innovation can redefine surface finishes, empowering industries to meet quality, performance, and sustainability demands in modern production.

REFERENCE

1. A. Bhardwaj, P. Ali, R. S. Walia, Q. Murtaza, and S. M. Pandey, "Development of Hybrid Forms of Abrasive Flow Machining Process: A Review," in *Advances in Industrial and Production Engineering*, K. Shanker, R. Shankar, and R. Sindhwani, Eds., in Lecture Notes in Mechanical Engineering. Singapore: Springer, 2019, pp. 41–67. doi: 10.1007/978–981-13–6412–9_5.
2. S. Mehta, P. Gauba, S. Kaushal, P. Ali, M. Dhanda, and R. S. Walia, "Developments in Hybrid Abrasive Flow Machining: A Review on Models and Analyses," in *Recent Trends in Mechanical Engineering*, A. Maurya, A. K. Srivastava, P. K. Jha, and S. M. Pandey, Eds., in Lecture Notes in Mechanical Engineering. Singapore: Springer Nature, 2023, pp. 375–384. doi: 10.1007/978–981-19–7709-1_38.
3. M. Ali-Tavoli, N. Nariman-Zadeh, A. Khakhali, and M. Mehran, "Multi-objective optimization of abrasive flow machining processes using polynomial neural networks and genetic algorithms," *Mach. Sci. Technol.*, vol. 10, no. 4, pp. 491–510, Dec. 2006, doi: 10.1080/10910340600996126.
4. A. W. Hashmi *et al.*, "Abrasive flow finishing of FDM printed extrusion die insert pattern using novel AFM fixture with mandrel guide," *Surf. Rev. Lett.*, vol. 30, p. 2350034, 2023, doi: https://doi.org/10.1142/S0218625X23500348.
5. R. Babbar and A. Misra, "Development of a novel magnetic abrasive finishing setup with replenishment of abrasive particles from external source and its experimental investigations," *CIRP J. Manuf. Sci. Technol.*, vol. 45, pp. 1–9, Oct. 2023, doi: 10.1016/j.cirpj.2023.06.001.
6. S. Ahmad, R. M. Singari, and R. S. Mishra, "Development of Al_2O_3-SiO_2 based magnetic abrasive by sintering method and its performance on Ti-6Al-4V during magnetic abrasive finishing," *Trans. IMF*, vol. 99, no. 2, pp. 94–101, Mar. 2021, doi: 10.1080/00202967.2021.1865644.

7. M. Anzai, K. Mazaki, and T. Nakagawa, "Development of magnetic abrasive using plasma powder melting method," *J Jpn Soc Abras Technol*, vol. 33, no. 4, pp. 33–38, 1989.
8. Y. Gao, Y. Zhao, and G. Zhang, "Preparation of Al_2O_3 magnetic abrasives by gas-solid two-phase double-stage atomization and rapid solidification," *Mater. Lett.*, vol. 215, pp. 300–304, Mar. 2018, doi: 10.1016/j.matlet.2017.12.124.
9. T. Y. Gridasova, A. F. Zhornyak, L. A. Karpova, G. G. Karyuk, V. E. Oliker, and V. Y. Shlyuko, "Magnetoabrasive materials from melts," *Sov. Powder Metall. Met. Ceram.*, vol. 19, pp. 505–507, 1980.
10. W. I. Kordonsky, I. V. Prokhorov, S. R. Gorodkin, G. R. Gorodkin, L. K. Gleb, and B. E. Kashevsky, "Magnetorheological polishing devices and methods," *Google Patents*, 1995.
11. F. Ji *et al.*, "The magnetorheological finishing (MRF) of potassium dihydrogen phosphate (KDP) crystal with Fe 3 O 4 nanoparticles," *Nanoscale Res. Lett.*, vol. 11, pp. 1–7, 2016.
12. S. Chen, S. Li, H. Hu, G. Tie, C. Guan, and Q. Li, "Analysis of surface quality and processing optimization of magnetorheological polishing of KDP crystal," *J. Opt.*, vol. 44, pp. 384–390, 2015.
13. J. Pan and Q. Yan, "Material removal mechanism of cluster magnetorheological effect in plane polishing," *Int. J. Adv. Manuf. Technol.*, vol. 81, pp. 2017–2026, 2015.
14. S. Han, F. Salvatore, J. Rech, and J. Bajolet, "Abrasive flow machining (AFM) finishing of conformal cooling channels created by selective laser melting (SLM)," *Precis. Eng.*, vol. 64, pp. 20–33, Jul. 2020, doi: 10.1016/j.precisioneng.2020.03.006.
15. S. Han, F. Salvatore, J. Rech, J. Bajolet, and J. Courbon, "Effect of abrasive flow machining (AFM) finish of selective laser melting (SLM) internal channels on fatigue performance," *J. Manuf. Process.*, vol. 59, pp. 248–257, Nov. 2020, doi: 10.1016/j.jmapro.2020.09.065.
16. R. Colaço and A. P. Serro, "Nanoscale wear of hard materials: An overview," *Curr. Opin. Colloid Interface Sci.*, vol. 47, pp. 118–125, Jun. 2020, doi: 10.1016/j.cocis.2020.01.001.
17. S. K. Everton, M. Hirsch, P. Stravroulakis, R. K. Leach, and A. T. Clare, "Review of in-situ process monitoring and in-situ metrology for metal additive manufacturing," *Mater. Des.*, vol. 95, pp. 431–445, Apr. 2016, doi: 10.1016/j.matdes.2016.01.099.
18. P. Ali, S. M. Pandey, M. S. Ranganath, R. S. Walia, and Q. Murtaza, "Experimentation and modelling of CNT additive abrasive media for micro finishing," *Measurement*, vol. 151, p. 107133, Feb. 2020, doi: 10.1016/j.measurement.2019.107133.
19. A. Hashmi, H. S. Mali, A. Meena, S. Ahmad, and Y. Tian, "A novel eco-friendly abrasive media based abrasive flow machining of 3D printed PLA parts using IGWO and ANN," *Rapid Prototyp. J.*, vol. 28, no. 1, pp. 1–20, 2023, doi: 10.1108/RPJ-04–2023–0136.
20. K. A. Khor, Z. H. Yuan, F. Y. C. Boey, Y. Murakoshi, and T. Sano, "Preparation of Al• Li/SiCp composite powder by a plasma spray atomisation (PSA) technique," *J. Mater. Process. Technol.*, vol. 48, no. 1–4, pp. 541–548, 1995.
21. V. K. Jain, "Magnetic field assisted abrasive based micro-/nano-finishing," *J. Mater. Process. Technol.*, vol. 209, no. 20, pp. 6022–6038, Nov. 2009, doi: 10.1016/j.jmatprotec.2009.08.015.
22. R. S. Walia, H. S. Shan, and P. Kumar, "Parametric optimization of Centrifugal Force-Assisted Abrasive Flow Machining (CFAAFM) by the Taguchi method," *Mater. Manuf. Process.*, vol. 21, no. 4, pp. 375–382, Jul. 2006, doi: 10.1080/10426910500411645.
23. R. S. Walia, H. S. Shan, and P. K. Kumar, "Finite element analysis of media used in the centrifugal force assisted abrasive flow machining process," *Proc. Inst. Mech. Eng. Part B J. Eng. Manuf.*, vol. 220, no. 11, pp. 1775–1785, Nov. 2006, doi: 10.1243/09544054JEM325.
24. A. Bhardwaj, K. Srinivas, and R. Chaudhary, "Analysis of shapes of centrifugal force-generating rod in centrifugal force-assisted abrasive flow machining process," *MAPAN*, vol. 38, no. 2, pp. 459–479, Jun. 2023, doi: 10.1007/s12647-023-00626-0.

25. M. R. Sankar, S. Mondal, J. Ramkumar, and V. K. Jain, "Experimental investigations and modeling of drill bit-guided abrasive flow finishing (DBG-AFF) process," *Int. J. Adv. Manuf. Technol.*, vol. 42, no. 7, pp. 678–688, Jun. 2009, doi: 10.1007/s00170-008-1642-y.
26. "A Robust Helical Abrasive Flow Machining (HLX-AFM) Process I SpringerLink." Accessed: Aug. 14, 2023. Online]. Available: https://link.springer.com/article/10.1007/s40032-012-0054-9
27. M. Ravi Sankar, V. K. Jain, and J. Ramkumar, "Rotational abrasive flow finishing (R-AFF) process and its effects on finished surface topography," *Int. J. Mach. Tools Manuf.*, vol. 50, no. 7, pp. 637–650, Jul. 2010, doi: 10.1016/j.ijmachtools.2010.03.007.
28. S. Singh and H. S. Shan, "Development of magneto abrasive flow machining process," *Int. J. Mach. Tools Manuf.*, vol. 42, no. 8, pp. 953–959, Jun. 2002, doi: 10.1016/S0890–6955(02)00021-4.
29. A. K. Sharma, G. Venkatesh, S. Rajesha, and P. Kumar, "Experimental investigations into ultrasonic-assisted abrasive flow machining (UAAFM) process," *Int. J. Adv. Manuf. Technol.*, vol. 80, no. 1, pp. 477–493, Sep. 2015, doi: 10.1007/s00170-015-7009-2.
30. B. S. Brar, R. S. Walia, and V. P. Singh, "Electrochemical-aided abrasive flow machining (ECA 2 FM) process: a hybrid machining process," *Int. J. Adv. Manuf. Technol.*, vol. 79, pp. 329–342, 2015.
31. R. Vaishya, R. S. Walia, and P. Kalra, "Design and development of hybrid electrochemical and centrifugal force assisted abrasive flow machining," *Mater. Today Proc.*, vol. 2, no. 4, pp. 3327–3341, Jan. 2015, doi: 10.1016/j.matpr.2015.07.158.
32. P. Ali, R. S. Walia, Q. Murtaza, and M. S. Ranganath, "Modeling and analysis of developed thermal additive Centrifugal Abrasive Flow Machining process," *Surf. Topogr. Metrol. Prop.*, vol. 8, no. 3, p. 035013, 2020.
33. P. Ali, R. S. Walia, Q. Murtaza, and Ranganath Muttanna Singari, "Material removal analysis of hybrid EDM-assisted centrifugal abrasive flow machining process for performance enhancement," *J. Braz. Soc. Mech. Sci. Eng.*, vol. 42, no. 6, p. 302, May 2020, doi: 10.1007/s40430-020-02375-6.
34. R. S. Walia, P. Ali, Q. Murtaza, and R. M. Singar, "Characterization of finished surface through thermal additive centrifugal abrasive flow machining for better surface integrity," *Indian J. Eng. Mater. Sci. IJEMS*, vol. 29, no. 1, pp. 29–44, 2022.
35. S. Ahmad, R. M. Singari, and R. S. Mishra, "Tri-objective constrained optimization of pulsating DC sourced magnetic abrasive finishing process parameters using artificial neural network and genetic algorithm," *Mater. Manuf. Process.*, pp. 1–15, Feb. 2021, doi: 10.1080/10426914.2020.1866196.
36. S. Ahmad, R. M. Singari, and R. S. Mishra, "Modelling and optimisation of magnetic abrasive finishing process based on a non-orthogonal array with ANN-GA approach," *Trans. IMF*, vol. 98, no. 4, pp. 186–198, Jul. 2020, doi: 10.1080/00202967.2020.1776966.
37. H. Yamaguchi and K. Hanada, "Development of spherical magnetic abrasive made by plasma spray," *J. Manuf. Sci. Eng.*, vol. 130, no. 3, pp. 031107-1–031107-9, Jun. 2008.
38. P. Singh, L. Singh, and S. Singh, "Manufacturing and performance analysis of mechanically alloyed magnetic abrasives for magneto abrasive flow finishing," *J. Manuf. Process.*, vol. 50, pp. 161–169, Feb. 2020, doi: 10.1016/j.jmapro.2019.12.033.
39. P. Kala, P. M. Pandey, G. C. Verma, and V. Sharma, "Understanding flexible abrasive brush behavior for double disk magnetic abrasive finishing based on force signature," *J. Manuf. Process.*, vol. 28, pp. 442–448, Aug. 2017, doi: 10.1016/j.jmapro.2017.04.010.
40. A. R. Chaudhari and K. B. Judal, "Experimental investigation of electro-chemical magnetic abrasive finishing of SS 304 workpiece," *Mater. Today Proc.*, vol. 49, pp. 390–396, 2022, doi: 10.1016/j.matpr.2021.02.295.
41. P. Kala and P. M. Pandey, "Experimental study on finishing forces in double disk magnetic abrasive finishing process while finishing paramagnetic workpiece," *Procedia Mater. Sci.*, vol. 5, pp. 1677–1684, 2014, doi: 10.1016/j.mspro.2014.07.356.

42. P. Kala, S. Kumar, and P. M. Pandey, "Polishing of copper alloy using double disk ultrasonic assisted magnetic abrasive polishing," *Mater. Manuf. Process.*, vol. 28, no. 2, pp. 200–206, Feb. 2013, doi: 10.1080/10426914.2012.746704.
43. R. S. Mulik and P. M. Pandey, "Ultrasonic assisted magnetic abrasive finishing of hardened AISI 52100 steel using unbonded SiC abrasives," *Int. J. Refract. Met. Hard Mater.*, vol. 29, no. 1, pp. 68–77, Jan. 2011, doi: 10.1016/j.ijrmhm.2010.08.002.
44. G.-W. Chang, B.-H. Yan, and R.-T. Hsu, "Study on cylindrical magnetic abrasive finishing using unbonded magnetic abrasives," *Int. J. Mach. Tools Manuf.*, vol. 42, no. 5, pp. 575–583, Apr. 2002, doi: 10.1016/S0890–6955(01)00153-5.
45. Y.-H. Lee, K.-L. Wu, J.-H. Jhou, Y.-H. Tsai, and B.-H. Yan, "Two-dimensional vibration-assisted magnetic abrasive finishing of stainless steel SUS304," *Int. J. Adv. Manuf. Technol.*, vol. 69, no. 9–12, pp. 2723–2733, Dec. 2013, doi: 10.1007/s00170-013-5242-0.
46. G. Y. Liu, Z. N. Guo, Y. B. Li, and J. W. Liu, "Composite tools design for electrolytic magnetic abrasive finishing process with FEM," *Adv. Mater. Res.*, vol. 325, pp. 536–541, 2011.
47. M. R. Muhamad, M. F. Jamaludin, M. S. A. Karim, F. Yusof, and Y. Zou, "Effects of electrolysis on magnetic abrasive finishing of AA6063-T1 tube internal surface using combination machining tool: Auswirkungen der Elektrolyse auf das Schleifen der Innenfläche eines AA6063-T1-Rohrs mit magnetischem Schleifmittel mit einem Kombinationswerkzeug," *Mater. Werkst.*, vol. 49, no. 4, pp. 442–452, Apr. 2018, doi: 10.1002/mawe.201700266.
48. T. S. Bedi and A. K. Singh, "An initial new approach for magnetorheological finishing of ferromagnetic internal cylindrical surfaces," *Int. J. Adv. Manuf. Technol.*, vol. 100, pp. 1017–1030, 2019.
49. Z. Sun, Z. Fan, Y. Tian, C. Qian, and Z. Ma, "Investigation on magnetorheological shear thickening finishing (MSTF) with radially slotted magnetic pole for free-form surface," *Int. J. Adv. Manuf. Technol.*, pp. 1–15, 2022.
50. J. D. Dent and T. E. Lang, "A biviscous modified bingham model of snow avalanche motion," *Ann. Glaciol.*, vol. 4, pp. 42–46, Jan. 1983, doi: 10.3189/S0260305500005218.
51. V. K. Jain, P. Ranjan, V. K. Suri, and R. Komanduri, "Chemo-mechanical magneto-rheological finishing (CMMRF) of silicon for microelectronics applications," *CIRP Ann.*, vol. 59, no. 1, pp. 323–328, 2010.
52. J. Tian, M. Chen, H. Liu, B. Qin, J. Cheng, and Y. Sun, "Study on mechanism of improving efficiency of permanent-magnet small ball-end magnetorheological polishing by increasing magnetorheological fluid temperature," *Sci. Rep.*, vol. 12, no. 1, Art. no. 1, May 2022, doi: 10.1038/s41598-022-11937-8.
53. F. Iqbal and S. Jha, "Experimental investigations into transient roughness reduction in ball-end magneto-rheological finishing process," *Mater. Manuf. Process.*, vol. 34, no. 2, pp. 224–231, 2019.
54. S. Jha and V. K. Jain, "Design and development of the magnetorheological abrasive flow finishing (MRAFF) process," *Int. J. Mach. Tools Manuf.*, vol. 44, no. 10, pp. 1019–1029, Aug. 2004, doi: 10.1016/j.ijmachtools.2004.03.007.
55. A. W. Hashmi *et al.*, "A comprehensive review on surface post-treatments for freeform surfaces of bio-implants," *J. Mater. Res. Technol.*, vol. 23, pp. 4866–4908, Mar. 2023, doi: 10.1016/j.jmrt.2023.02.007.
56. F. Shi, Y. Tian, X. Peng, and Y. Dai, "Combined technique of elastic magnetorheological finishing and HF etching for high-efficiency improving of the laser-induced damage threshold of fused silica optics," *Appl. Opt.*, vol. 53, no. 4, pp. 598–604, 2014.
57. G. Ghosh, A. Sidpara, and P. P. Bandyopadhyay, "Performance improvement of magnetorheological finishing using chemical etchant and diamond-graphene based magnetic abrasives," *Precis. Eng.*, vol. 79, pp. 221–235, 2023.

58. S. Ahmad, R. K. Singh, R. M. Singari, and R. Mishra, "On the Relationship Between Surface Microhardness and Roughness Produced by MAF Process," in *Processing and Characterization of Materials*, S. Pal, D. Roy, and S. K. Sinha, Eds., Springer, 2021, pp. 9–24.
59. Z. Q. Liu, Y. Chen, Y. J. Li, and X. Zhang, "Comprehensive performance evaluation of the magnetic abrasive particles," *Int. J. Adv. Manuf. Technol.*, vol. 68, no. 1–4, pp. 631–640, Sep. 2013, doi: 10.1007/s00170-013-4783–6.
60. B. Yang, W. Lu, W. Feng, X. Yang, and D. Zuo, "Adsorption and deposition of micro diamond particles in preparing diamond magnetic abrasives by electroless composite plating," *Diam. Relat. Mater.*, vol. 73, pp. 137–142, Mar. 2017, doi: 10.1016/j.diamond.2016.08.015.
61. W. Li, X. Li, S. Yang, and W. Li, "A newly developed media for magnetic abrasive finishing process: Material removal behavior and finishing performance," *J. Mater. Process. Technol.*, vol. 260, pp. 20–29, Oct. 2018, doi: 10.1016/j.jmatprotec.2018.05.007.
62. A. C. Wang and S. J. Lee, "Study the characteristics of magnetic finishing with gel abrasive," *Int. J. Mach. Tools Manuf.*, vol. 49, no. 14, pp. 1063–1069, Nov. 2009, doi: 10.1016/j.ijmachtools.2009.07.009.
63. W. Yao, B. Lyu, T. Zhang, L. Guo, and Y. Song, "Effect of elastohydrodynamic characteristics on surface roughness in cylindrical shear thickening polishing process," *Wear*, p. 205026, 2023.
64. A. Sidpara and V. K. Jain, "Effect of fluid composition on nanofinishing of single-crystal silicon by magnetic field-assisted finishing process," *Int. J. Adv. Manuf. Technol.*, vol. 55, no. 1–4, pp. 243–252, Jul. 2011, doi: 10.1007/s00170-010-3032–5.
65. A. Sidpara, M. Das, and V. K. Jain, "Rheological characterization of magnetorheological finishing fluid," *Mater. Manuf. Process.*, vol. 24, no. 12, pp. 1467–1478, 2009.
66. G. Liu *et al.*, "Preparation of Al_2O_3 magnetic abrasives by combining plasma molten metal powder with sprayed abrasive powder," *Ceram. Int.*, vol. 48, no. 15, pp. 21571–21578, Aug. 2022, doi: 10.1016/j.ceramint.2022.04.129.
67. M. Feng *et al.*, "Effect of the components of Magnetic Compound Fluid (MCF) slurry on polishing characteristics in aspheric-surface finishing with the doughnut-shaped MCF tool," *Precis. Eng.*, vol. 65, pp. 216–229, 2020.
68. N. Dixit, V. Sharma, and P. Kumar, "Experimental investigations into abrasive flow machining (AFM) of 3D printed ABS and PLA parts," *Rapid Prototyp. J.*, vol. 28, no. 1, pp. 161–174, 2022.
69. Z. Fan, Y. Tian, Q. Zhou, and C. Shi, "Enhanced magnetic abrasive finishing of Ti–6Al–4V using shear thickening fluids additives," *Precis. Eng.*, vol. 64, pp. 300–306, Jul. 2020, doi: 10.1016/j.precisioneng.2020.05.001.
70. Z. Ma, Y. Tian, C. Qian, S. Ahmad, Z. Fan, and Z. Sun, "Modeling and simulation of material removal characteristics in magnetorheological shear thickening polishing," *Int. J. Adv. Manuf. Technol.*, vol. 128, no. 5, pp. 2319–2331, Sep. 2023, doi: 10.1007/s00170-023-12093-7.
71. X. Li, Q. Yu, X. Zhou, Y. Zhang, R. Lv, and Y. Zhao, "Magnetic sensing technology of fiber optic interferometer based on magnetic fluid: A review," *Measurement*, vol. 216, p. 112929, Jul. 2023, doi: 10.1016/j.measurement.2023.112929.
72. F. W. Preston, "The theory and design of plate glass polishing machines," *J. Glass Technol.*, vol. 11, no. 44, pp. 214–256, 1927.
73. J. Deng, Q. Zhang, J. Lu, Q. Yan, J. Pan, and R. Chen, "Prediction of the surface roughness and material removal rate in chemical mechanical polishing of single-crystal SiC via a back-propagation neural network," *Precis. Eng.*, vol. 72, pp. 102–110, Nov. 2021, doi: 10.1016/j.precisioneng.2021.04.012.
74. T.-T. Do and T.-H. Fang, "Deep insights into interaction behaviour and material removal of β-SiC wafer in nanoscale polishing," *Tribol. Int.*, vol. 186, p. 108639, 2023.

75. M. Bhagavat, F. Yang, and I. Kao, "Elasto-plastic finite element analysis of indentations in free abrasive machining," in *ASME International Mechanical Engineering Congress and Exposition*, American Society of Mechanical Engineers, 1998, pp. 819–824.
76. T. Wang, D. Tan, W. Xu, C. Wang, Y. Tan, and Y. Hou, "Investigation on abrasive-wall collision mechanism and the universal design method for constraint module in soft abrasive flow polishing," *Int. J. Adv. Manuf. Technol.*, pp. 1–16, 2023.
77. S. Mohseni-Mofidi, L. Pastewka, M. Teschner, and C. Bierwisch, "Magnetic-assisted soft abrasive flow machining studied with smoothed particle hydrodynamics," *Appl. Math. Model.*, vol. 101, pp. 38–54, Jan. 2022, doi: 10.1016/j.apm.2021.07.015.
78. T. Hemphill, W. Campos, and A. Pilehvari, "Yield-power law model more accurately predicts mud rheology," *Oil Gas J. US*, vol. 91, no. 34, 1993.
79. A. Shahzad and A. L. I. Ramzan, "MHD flow of a non-Newtonian Power law fluid over a vertical stretching sheet with the convective boundary condition," *Walailak J. Sci. Technol. WJST*, vol. 10, no. 1, pp. 43–56, 2013.
80. M. Arif, P. Kumam, W. Kumam, I. Khan, and M. Ramzan, "A fractional model of Casson fluid with ramped wall temperature: engineering applications of engine oil," *Comput. Math. Methods*, vol. 3, no. 6, p. e1162, 2021.
81. E. A. Kirsanov and S. V. Remizov, "Application of the Casson model to thixotropic waxy crude oil," *Rheol. Acta*, vol. 38, no. 2, pp. 172–176, 1999.
82. X. Huang and M. H. Garcia, "A Herschel–Bulkley model for mud flow down a slope," *J. Fluid Mech.*, vol. 374, pp. 305–333, 1998.
83. D. Wang, G. Barakos, Z. Cheng, H. Mischo, and J. Zhao, "Numerical simulation of pressure profile of mining backfill fly-ash slurry in an L-shaped pipe using a validated Herschel-Bulkley model," *J. Sustain. Cem.-Based Mater.*, vol. 12, no. 3, pp. 219–233, 2023.
84. A. Malpotra, B. Singh, and L. Singh, "Electrolytic magnetic abrasive finishing process – A review," *Mater. Today Proc.*, Mar. 2023, doi: 10.1016/j.matpr.2023.03.237.
85. Y. Choopani, M. R. Razfar, M. Khajehzadeh, and M. Khosrojerdi, "Design and development of ultrasonic assisted-rotational magnetorheological abrasive flow finishing (UA-RMRAFF) process," *Appl. Acoust.*, vol. 197, p. 108950, Aug. 2022, doi: 10.1016/j.apacoust.2022.108950.
86. J. Cao *et al.*, "A novel surface polishing method and its fundamental performance in ultra-fine polishing of wafer," *Int. J. Adv. Manuf. Technol.*, vol. 105, pp. 2919–2933, 2019.

Index